Low Power and Low Voltage Circuit Design with the FGMOS Transistor

Other volumes in the Circuits, Devices and Systems series:

Low Power and Low Voltage Circuit Design with the FGMOS Transistor

Dr Esther Rodriguez-Villegas

The Institution of Engineering and Technology

Published by The Institution of Engineering and Technology, London,
United Kingdom

© 2006 The Institution of Engineering and Technology

First Published 2006

The Institution of Engineering and Technology,
Michael Faraday House,
Six Hills Way, Stevenage,
Herts, SG1 2AY, United Kingdom

www.theiet.org

British Library Cataloguing in Publication Data

Rodriguez-Villegas, Esther
    Low power and low voltage circuit design with the FGMOS
    transistor. — (IEE circuits, devices and systems series)
    1. Power transistors 2. Low voltage integrated circuits
    I. Title II. Institution of Engineering and Technology

621.3'81528

ISBN (10 digit) 0 86341 617 9
ISBN (13 digit) 978-086341-617-0

Typeset in India by Newgen Imaging Systems (P) Ltd, Chennai, India
Printed in the UK by MPG Books Ltd, Bodmin, Cornwall

For my parents:
Jesus and Maria Josefa

Contents

Preface

The implantation of electronic devices into biological systems is an exciting new development in medical science. Implantable prosthesis has the potential to provide medical practitioners with a range of new diagnostic tools and treatment options. These devices must be small and lightweight, which means their batteries must be small in number and be long lasting. From the circuit design point of view this means low voltage and low power (LV/LP). At the same time consumers are demanding greater portability and longer battery life from their handheld and laptop devices. Again, to the circuit designer this means LV/LP.

Partly due to the demand for smaller and faster products, there is an ongoing trend in fabrication processes towards smaller transistors. Future deep sub-micron devices will not be able to withstand high operating voltages, so circuit design needs to reflect this fact. The SIA predicts that in 2010 the maximum supply voltage will be 0.8 V[1]. The current twin aims of circuit design research are low voltage and low power. The former is required to meet the requirements of sub-micron fabrication processes and the latter is required to meet battery longevity requirements. In digital circuits power consumption is proportional to V_{DD}^2, where V_{DD} is the supply voltage. Most current and future designs are likely to use the system-on-a-chip (SOC) design style. SOC integration means that analog and digital circuits must use the same process and share a common substrate. Whilst the performance of digital circuits is relatively robust to transistor shrinkage, the performance of analog circuits generally degrades because making the transistors smaller necessitates a reduction in V_{DD}.

The drive towards very low voltage analog circuits did not gather momentum until the mid-nineties. As a consequence, the design techniques are not well developed. One of the most promising strategies in LV/LP analog circuit design is the floating gate MOS (FGMOS) transistor. These devices can be fabricated in all CMOS technologies, although a double poly CMOS technology is preferred. FGMOS allows the designer to do any or all of the following: reduce the complexity of circuits; simplify the signal processing chain within a design; shift the signal levels and incorporate

---

[1] The National Technology Roadmap for Semiconductors. 2002 Update. Semiconductor Industry Association, 2003.

tuneable mechanisms. These new possibilities make it feasible to design circuits with functionality that would not be possible as supply voltages get lower in comparison to the threshold voltages of specific technologies.

This book is intended for a broad range of readers. Although the title seems to suggest that it is a book on the FGMOS and therefore only those interested in exploring this area can benefit from it, the reality is different. This is fundamentally a book on circuit design pushing the limits of CMOS technologies. As the power and voltage constraints become more and more restrictive the performance of conventional circuit topologies degrades up to the point that they stop being functional. The tradeoffs need to be established much more exhaustively and "rules of thumb" no longer help. In some cases alternative circuit techniques have to be found. This book guides the reader through a thorough analysis of design tradeoffs for every circuit topology presented in it together with a quantitative and qualitative study of what the consequences of shrinking the voltage levels and reducing power are. From this point of view, it is a book that can appeal to any audience with an interest in low power and low voltage circuit design.

But also, it is a book on the FGMOS transistor and its use in analog and digital circuits. The text presents the device from scratch and builds up from there so no previous knowledge is needed. Hence, a designer with previous knowledge in the subject can probably skip parts of chapters two and three. As Section 1.5 explains the book is mostly focused on analog because it is in this context that the FGMOS transistor has more potential. Nevertheless some very interesting applications can also be found in the digital area and this is why the last chapter is on low power and low voltage digital design. Unlike the analog chapters which are much more incremental, this chapter requires a previous basic digital knowledge.

Acknowledgments

Part of the material presented in this book is the result of the research carried out in the Seville Institute of Microelectronics (IMSE-CNM) and was sponsored by the Spanish government under an FPU grant. This sponsorship is gratefully acknowledged.

I am also thankful to many of the people working there at the time specially to my then Ph.D. supervisors Dr. Alberto Yufera Garcia and Professor Adoracion Rueda Rueda. The material presented in chapter number 8 was gathered with the help of Dr. Jose Maria Quintana and Dr. Maria Jose Avedillo who set up the simulation environments for the traditional implementation of the digital circuits.

Apart from those who supported me on the technical and financial side, there were many others who played an even more important role, lifting up my mood on many occasions I just wanted to give up electronics and become a "chocolate critic". Thank you to my "friends in the Himalaya": Kike, Carlos, Antonio, Pedro, Hakim and specially Africa (thank you for being such a wonderful friend through all these years). Also to many others in the CNM, mostly Eduardo and Emi. Thank you to my "friends in Circuits and Systems at Imperial": Ganesh, Anjit, Solon, Thomas, Rohit, David and Omeni (I miss you now that I do not see you everyday). And those in "Signal Processing": Jon, Thushara, Alex, David and Hugh. Thank you to Fernando for putting up with me and the FGMOS when no one believed my "getting rid of the charge" technique was going to work.

Also, thank you to my colleagues at Imperial College especially to Professor Peter Cheung, Dr. David Haigh and Dr. Christos Pappavassiliou.

Thank you to Dr. Ramon Gonzalez-Carvajal for believing in me, helping me every time I asked for help (and many other times he could read my mind), making me laugh and showing me the best tapas bars in Seville.

Very special thank you to Mr. Phil Corbishley for being such a good student, friend and for reading throughout the whole text.

Thank you to Dr. Patrick Degenaar since without him this book would not have been completed.

Thank you to Dr. Aleksandra Rankov for helping me with the edition.

Thank you to Mr. Vasa Curcin for being so sweet, being always there and dragging me out for coffee when I needed it the most.

Thank you to all the people in the IEE who have worked together on publishing this book.

And finally, thank you to my family (specially my sisters, Maria and Bea), who lifted me in very difficult moments.

Chapter 1

Introduction

1.1 The general framework

1.1.1 Why analog? Why digital?

In recent years, digital signal processing has progressively supplanted analog signal processing in chip design. This is due to it having lower development costs, better precision performance and dynamic range, as well as being easier to test. The role of analog circuits has been mostly restricted to electronic applications of interfacing digital systems to the external world. Nevertheless, when precise computation of numbers is not required (as is the case in systems designed for perception of a continuously changing environment), and massively parallel collective processing of signals is needed, low precision analog VLSI (very large scale integration) has proven to be more convenient than digital in terms of cost, size and/or power consumption [1].

In recent years mixed signal application-specific integrated circuits (ASICs) have become increasingly popular. The cooperative coexistence of analog and digital circuits is very beneficial since they compensate for each other's weaknesses. Hence, although in many aspects digital electronics is superior, in reality it requires a symbiotic relationship with analog.

1.1.2 Why low voltage?

Since the invention of the transistor more than 50 years ago, the progress of microelectronics can be summarised as follows: 15 per cent decrease in feature size per year, 30 per cent cost decrease per year, 50 per cent performance improvement and 15 per cent semiconductor market growth rate. The numbers speak for themselves. This exponential evolution made many experts in the 1990s assert that fundamental limits were about to be reached. Fortunately, technical innovations made it possible to shrink the technologies to smaller dimensions than the predicted 0.3 μm. However, as the dimensions of the devices reduce, a new constraint arises: the interconnect delays and the fact that they directly affect the CV^2 power dissipation. In the past, this was not a problem as the capacitances were scaled down together with the dimensions. Recently this scaling relationship has been replaced to being proportional to the total length of wires, L, in the circuit. The interconnect power dissipation can therefore be rewritten as $kV^2 L$ (where k is the dielectric permittivity). Hence, the most significant parameter in the reduction of the interconnect power is the voltage and new strategies are required to operate circuits at lower power supply voltages [2].

However, this is not the only motivation fuelling the eagerness of researchers to operate circuits at lower voltages. The other one is related to the magnitude of the electric fields in the devices. These grow proportionally as the dimensions are scaled down, which increases the risk of dielectric breakdown. This can additionally be compensated for by reducing the voltage differences across the devices. Hence a low voltage power supply is beneficial.

1.1.3   Why low power?

The fast development of electronic-based entertainment, computing and communication tools, especially portable ones, has provided a strong technology drive for microelectronics during the last ten years. System portability usually requires battery supply and therefore weight/energy storage considerations. Unfortunately, battery technologies do not evolve as fast as the applications demand. Therefore the challenge, derived from market requirements, is to reduce the power consumption of the circuits.

In addition to consumer products, battery lifetime is a crucial factor in some biomedical products which have to be either worn or implanted within the patients for a long period of time; such systems are continuously increasing in number and in scope. Investigation into low power biomedical systems is another interesting quest for microelectronic designers [3–6].

1.1.4   Why CMOS?

The choice of fully integrated VLSI complementary metal oxide semiconductor (CMOS) implementations is based on their lower cost (whenever the state-of-the-art sub-micron VLSI processes are not required) and design portability. Furthermore, CMOS technologies allow the possibility of integration with micro electro mechanical systems (MEMS) [7]. These are the most important reasons that have directed the semiconductor industry towards CMOS mixed signal designs, and place CMOS technologies as the leader in the microelectronics semiconductor industry [8].

1.2   Techniques to reduce the power consumption and the voltage supply

1.2.1   Analog techniques

Different techniques have been developed to reduce power consumption and supply voltage in analog circuits. They can be classified into two main groups: techniques intended to reduce the voltage levels and techniques intended to reduce the current levels.

1.2.1.1 Techniques for voltage reduction

The most popular techniques for voltage reduction are as follows:

1. *Circuits with a rail-to-rail operating range*: This group includes all the techniques that are meant to extend the voltage ranges of the signals. Most of these techniques are based on redesigning the input and output stages in order to increase their linear range [9–13]. In these topologies the transistors have to be biased in those regions that optimise the operating range. Since the voltage constraints are more restrictive, in order to get the devices working in a certain region, sometimes it is necessary to shift the voltage levels. In this book the reader will find multiple examples of how the floating gate MOS (FGMOS) can be used to shift the voltage levels in a very simple way, so the operating range is optimised.

2. *Technique of cascading stages, instead of a single cascode stage*: Conventional circuit topologies stacked cascode transistors in order to obtain the high output resistances and gains required by certain structures such as OPAMPs and OTAs. However, stacking transistors in a given circuit branch makes the voltage requirements more demanding for the entire cell. The solution for this is either to reduce the voltage requirements for the transistors, or to substitute each single stage for a cascade of them, in such a way that the total gain at the output would be the product of all the single ones (whenever high gain is needed). If the latter route is taken, there is an added problem related to frequency stabilisation [14–16]. Furthermore, as the number of branches increases, so too does the power consumption. Therefore a compromise solution would be to still use cascode transistors, but try to minimise their voltage requirements within the whole operating range. This book will show how to obtain this by using cascode devices implemented with FGMOS transistors. Different circuits will be presented, that, even with a single stage, achieve a high performance in terms of gain under a very low power supply voltage.

3. *Supply multipliers*: Charge pumps can be used to scale up the power supply voltage for certain analog cells, while still keeping the low supply voltage value for the digital blocks [17–20]. The main drawback of this technique is that it requires large capacitors which take a large silicon area, a considerable overhead for circuits using this technique. In addition, the extra power consumption can be considerable.

4. *Nonlinear processing of the signals*: Most practical electronic systems are designed to process signals in a linear form. However, the fact that the input/output relationship between two variables has to be linear does not mean that internally the system must be linear as well [21]. The fundamental devices constituting the blocks are transistors which are inherently nonlinear. Traditional circuit techniques tried to linearise the behavioural laws of the devices with more or less complicated topologic solutions that in most cases unavoidably increased the power consumption. The idea behind nonlinear techniques is to exploit the nonlinear I/V characteristics of the transistors to process the signals more efficiently. A very powerful method to obtain this consists of processing the input signal in such a way that for low input values

the internal signal levels are still above the maximum expected noise, whereas large input signals are attenuated, hence reducing distortion. Ideally, a function would be chosen that shrinks the internal dynamic range (DR) down to 0 dB and once the signal is internally processed, the inverse is applied to recover the original DR expansion. This would keep the DR, as well as (a) reduce the large voltage swings around the bias point, hence relaxing the voltage constraints imposed by the lower power supply voltage and (b) reduce the amount of power needed to charge and discharge the parasitic capacitances.

The 'compressing/expanding' processing technique at the device level is known as instantaneous companding. Other techniques at the system level are for example syllabic companding [22–24] and time-variant compressing techniques for AGC [25].

The majority of the currently existing companding circuits base their functionality on the nonlinear current law of the bipolar transistor, and they are not compatible with the low power and low voltage constraints since they were originally developed to exploit the device's high frequency capability. Unfortunately these techniques are not easy to translate to a MOS implementation, because of several reasons such as asymmetric I/V curves in MOS devices and mismatch. On top of that, most of the existing CMOS realisations cannot be generalised due to, for example, local bulks, poor low voltage operation or redundant circuitry [26–29].

In this book, the reader will find how to overcome some of the problems inherent to MOS realisations by using FGMOS transistors to realise externally linear, internally nonlinear circuits. Besides, complex functionality can be implemented with very compact designs which is also beneficial from the point of view of power.

1.2.1.2   Techniques for current reduction

The main circuit design techniques oriented towards reduction of current are

1. *Adaptive biasing:* This technique is based on using a non-static current bias to optimise the power consumption according to signal demands [30,31].
2. *Subthreshold biasing:* Another way to reduce the current levels and hence the power is by using MOS transistors biased in the weak inversion region driving very low current levels. Some sections of the book will show how by employing FGMOS transistors instead an extra power reduction can be achieved, thanks to its better suitability to implementation of very complex functions [32].

1.2.2   Digital techniques

Power dissipation in CMOS digital circuits can be classified into static power consumption, which is the consequence of the resistive paths from the power supply to ground; and dynamic power consumption, which is due to the switching of capacitive loads between two different voltage states.

Until relatively recently, digital designers optimised their designs to meet performance and area constraint; power was a small detail to think about after all the other requirements. The most important low power design method was simply to avoid 'wasting' it, and if power was a target obvious effective solutions were either

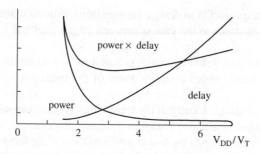

Figure 1.1   Power and delay versus power supply voltage measured in threshold voltage units [33]

just to reduce the power supply voltage, remove circuits that dissipate static power or to power down inactive blocks. But using power as a metric is not completely appropriate in the context of digital design, as power can always be reduced by slowing down the operating frequency. A more suitable metric is the figure of merit delay × power. Smaller delay × power means a lower energy solution at the same level of performance. The most popular techniques to improve digital circuits performance according to this metric can be classified as follows [33]:

1. *Voltage scaling:* A reduction of the power supply voltage (V_{DD}) brings with it an immediate reduction of the energy per operation. But, as the capacitance (C) and the threshold voltage (V_T) are constant, the speed of the gates[2] will also decrease in agreement with eq. (1.1) [34]:

$$t_d \propto \frac{CV_{DD}}{(V_{DD} - V_T)^2} \tag{1.1}$$

As for the power, whenever the dynamic term is dominant, it can be written as

$$P \propto C \cdot V_{DD}^2 f \tag{1.2}$$

These equations show that for V_{DD} values much greater than V_T, small decreases in V_{DD} lead to small increases in delay but much larger decreases in power consumption. However, as V_{DD} approaches the threshold voltage, further decreases in V_{DD} lead to much greater increases in delay for the equivalent decrease in power. Finally, the variations are not significant for V_{DD} values larger than three times the threshold voltage. Around this point the figure of merit remains almost constant. These tradeoffs are graphically illustrated in Fig. 1.1.

2. *Transistor sizing:* The main contribution to the load capacitance is the gate capacitance of the forthcoming gate, so by making these capacitances smaller the power consumption will also decrease. However, for a minimum transistor length, a reduction on its width also reduces the maximum value of the current it can drive.

---

[2] t_d is the delay time. f is the operating frequency.

In general, the best tradeoff is to design the transistors with such dimensions that the percentage of the load due to the gate is between 20 per cent and 80 per cent of the total.

3. *Adiabatic circuits:* This approach is based on resonating the loading capacitance with an inductor which recovers some of the energy loss in switching the load [35].

4. *Technology scaling:* Ideally, if the technology is scaled in such a way that all voltages and dimensions are also scaled in the same proportion, the delay × power product would be reduced up to the fourth power of the scaling factor. But, this would only be true if the static power is negligible. Otherwise it would just improve as the square of it. However, there is a limit for this determined by the interconnections.

5. *Transition reduction:* This technique consists of selectively powering down those sections that do not have any influence on the circuit response during a certain period of time. The activation is usually done by gating the clock to the function block.

6. *Operand isolation:* This technique is applicable when the circuit inputs are changing rapidly but the output is only needed once every few clock cycles. In this case, the input signal can be isolated from the rest of the circuit, so reducing switching and thereby the power consumption.

7. *Precomputation:* This technique consists of breaking the function performed on a large bus of data into two parts: a small and a large block. The small block is precomputed and power is consumed in the rest of the function only if it is valid.

8. *Parallelism:* This technique consists of having a number N of identical functional units operating in parallel. This increases the energy per operation by only the power overhead while the delay per operation drops by N minus the delay overhead.

9. *Problem redefinition:* The most effective way of reducing the power consumption in digital systems is to reduce the number/complexity of tasks that the operation requires. It is at this level that the designer has a bigger task, as simplifications often reduce both the energy and the delay to complete the operation. A reformulation of the problem can lead to a solution that requires less computation to accomplish the same task – it is possible to obtain a 50 per cent power reduction from architectural changes.

The last chapter of this book shows how it is possible to develop digital circuits topologies that considerably reduce the figure of merit power × delay by using the FGMOS transistor as a computational block.

1.3   Why floating gate MOS?

The aim of this book is to show the reader how to use a device that can be fabricated in all CMOS technologies, the FGMOS, to design circuits that (a) can operate at power supply voltage levels which are well below the intended operational limits for a particular technology and (b) consume less power than the minimum required power of a circuit designed with only MOS devices in the same technology with the same performance.

The two main design targets will be low voltage and low power (LV/LP). They will be achieved by pursuing four different subgoals.

1. *Reduce the circuit complexity:* As the circuitry gets simpler (fewer devices), fewer current branches are required, and therefore the power consumption decreases. This also has other benefits related to the frequency response, since the number of internal nodes is smaller.
2. *Simplify the signal processing:* Complex functions are easier to implement using FGMOS transistors. These will be used in nonlinear signal processing to reduce the voltage demands.
3. *Shift the signal levels:* The devices will be biased in the most appropriate operating region for a wider range of input signals, by shifting the effective threshold voltages accordingly in the FGMOS transistors. This can be achieved without the need for extra level shifters, although in some cases it can be detrimental as well. These cases will also be discussed throughout the book.
4. *Facilitate the tuning:* Tuning becomes even more of an issue in a low voltage context, where variations are more critical since they can bias the devices out of their intended operating region. FGMOS transistors increase the number of degrees of freedom available to tune/program the circuits.

The book will present the device as a powerful mathematic/electronic element, which offers three very important properties in the LV/LP context:

Flexibility, Controllability, Tunability

Flexibility to implement both linear as well as complex nonlinear functions in a compact and easy fashion, thus leading to the consequent simplification of the circuits.
Controllability, since the effective threshold voltage of every single transistor can be controlled separately according to the operating range needs.
Tunability, because it is a multiple input device and can be designed to be tuned just by adding extra inputs.

These properties will be extensively illustrated throughout this book as they will be used in the design of several LV/LP analog and digital blocks.

1.4   FGMOS history

Due to the special characteristics of the FGMOS transistor, its application in both analog and digital circuits has been very wide since the first report in 1967 [36,37]. The first well-known application of the FGMOS was to store data in EEPROMs, EPROMs and FLASH memories. In the late 1980s, the Intel ETANN chip, employed it as an analog nonvolatile memory element [38]. Today this technology is present in every personal computer [39]. But this was not the only context the transistor has been used in. During the last ten years, a number of different applications have revealed possibilities that this device could have in many other different fields. This results from its versatility in implementing different functions, as well as its capacity for programmability. In the following text, several examples illustrating use of the device in different frameworks will be cited, together with references where many others can be found.

The FGMOS has been widely used as a trimming element for tuning purposes. An example of this can be found in Sackinger and Guggenbuhl's work [40]. They proposed an application for the FGMOS transistors as tunable elements in analog CMOS amplifiers. The idea was to use these tunable elements to correct offset errors, linearity, gain errors and so on. Previously, this was done using laser techniques, programming of resistor networks or dynamic compensation. It was then proposed that the correction process be carried out by means of a continuous modification in the accumulated charge at the floating gate. The tuning was done by means of electric pulses, thus avoiding the need for expensive laser equipment and preventing heating. Another advantage was, that unlike previous works on nonvolatile memories, in which the variations in the threshold voltages could be around 40 per cent in four days, if FGMOS transistors were used these changes could be much smaller (around 1 per cent in ten years at room temperature, in a process designed for digital memories). Moreover, tunable circuits implemented with floating gates could be reprogrammed many times, making this technique good not only for tuning objectives but also for synapses, either in neural networks or in adaptive filters.

Yu and Geiger [41] suggested a similar method to design analog circuits working at a very low voltage supply, scaling down the threshold voltage and also injecting charge using the tunnel effect [42], but with the difference that the high voltages were not obtained externally but internally with a charge pump, itself designed with floating gates. In order to control the charge at the gate, Fowler–Nordheim tunnelling current had to be generated. This required an EEPROM technology consisting of a standard CMOS technology with an extension in the process to have an ultra-thin oxide so that the tunnel effect could be forced, transistors which could bear high voltages and a second polysilicon layer. All these requirements contributed to the increase in the chip cost. Besides, these technologies were generally lagging one or two generations behind the state-of-the-art CMOS technologies. This implied a penalty in the integration density and also, due to the rather small size of the market for EEPROM technologies, an effective monopoly is in place. Therefore, a circuit with an on-chip EEPROM is expensive. The cost argument is especially valid for ICs where the EEPROM is only a small fraction of the circuit.

Research focusing on overcoming this drawback led Thomsen *et al.* to fabricate a tunnelling injector in a standard double-polysilicon CMOS process [43]. Op't Eynde and Zorio [44] proposed a new alternative for using FGMOS in circuits which only had to memorise a few tens of bits, e.g. for circuit identification or for the trimming of analog circuits by means of digital calibration. The EEPROM function was realised in a standard CMOS technology, even when the resulting cell was larger than a normal EEPROM. The use of even higher voltages was essential to provoke tunnelling (around 35 V). However, in order to avoid the dielectric breakdown, time-dependent-dielectric-breakdown (TDDB) or gate-induced-drain-leakage (GIDL) these voltages were never applied directly to the transistor. The handicap was that these memories were useful for applications where only a few write-and-erase cycles were required and the high programming voltage was applied externally during programming. Other examples where the FGMOS transistor was used as a trimming element with the help of the tunnel effect can be found from [45] to [49].

Shibata and Ohmi [50–53] proposed a different application for the transistor and a different name: Neuron MOS (νMOS), since its functional behaviour is quite analogous to that of biological neurons: the 'on' and 'off' of the transistor is based on the result of a weighted sum operation. The transistor was used in multivalued logic [54–57], with the advantages this had over binary logic, since a large amount of data can be processed per unit of area, with a reduced number of connections. This was proved with NMOS and CMOS implementations of the hardware algorithms. Generally, although the calculations were implemented in current mode increasing the power dissipation, the νMOS realises the sum operation in the voltage mode using a capacitive division, and so the current flow is zero with the exception of capacitance charges and discharges. This was very important, when the integration density was very large [58].

Simultaneously to Shibata and Ohmi's work, a different research was carried out, in which the FGMOS transistor was used as a memory element in neural networks. But it was still difficult to control the analog data stored at the gate with a high resolution. The fine tuning of the charge injected by tunnelling was not easy, as the tunnelling current did not have a linear dependence with the junction voltage. Feedback circuits were an option but this would increment the chip area and the power consumption. Fujita and Amamiya [59] developed an FGMOS device which could be used as a precision analog memory for neural networks LSI (large scale integration). This device had two floating gates. One was a charge-injection gate with a Fowler–Nordheim tunnel junction and the other was a charge-storage gate that operated as an MOS floating gate. The gates were connected by a high resistance, and the charge-injection gate was small so that its capacitance was much smaller than that of the charge-storage gate. By applying control pulses to the charge-injection gate, it was possible to charge and to discharge the MOS floating gate in order to modify the current with high resolution over 10 bits. This had a high applicability for on-chip learning in analog neural network LSIs.

Learning on a neural network is the change in the synaptic weight according to the learning algorithm. The use of the floating gate EEPROM technology for electronic synapses has presented a crucial problem: the amount of incremental data change by a constant programming pulse is not proportional to the number of pulses. The reason is that the electric field in the tunnel oxide is determined not only by the magnitude of a programming pulse but also by the amount of charge stored in the floating gate. The strong nonlinear dependence on the electric field made it extremely difficult to update the weight accurately via the number of pulses. Thus, to achieve this, complicated control circuitry was usually employed [60]. Kosaka *et al.* [61,62] presented a synapse with a good weight updating linearity under constant pulse programming. This was realised by employing a simple self-feedback regime in each cell circuit. As it was very simple, neither the speed nor the integration density were affected. It is also worth noting the work of P. Hasler, C. Diorio and B. Minch, which will be delved into deeper in the forthcoming text. In any case, a more in-depth analysis of the FGMOS evolution in the adaptive circuits context can be carried out by reading other relevant research [63–81].

In addition to the memory, neural networks and logic systems applications above, the FGMOS has other potential possibilities. These result from the consequence of the weighted sum of voltages that can be performed in a lossless node. This allows a simplification in many blocks for analog signal processing. Yang and Andreou proved the effectiveness of the technique by designing a multiple input differential amplifier which functioned like a standard differential difference amplifier (DDA). It had some unique advantages, including lower circuit complexity and better input transistor matching [82], but the accumulated charge at the floating gate was not altered. The possible residual charge that might remain trapped at the gate during the fabrication process was an unknown parameter which had to be eliminated. They accomplished this using ultraviolet light [83–85]. These authors also developed a model to design with FGMOS transistors working in the subthreshold region [86].

The model was then used by Hasler, Minch, Diorio and Mead's group to carry out their work based on the FGMOS transistor in subthreshold, which comprised several applications for this transistor [87] – from nonlinear functions implementation in current mode [88,89] to the design of memory cells [90,91], combining the tunnel effect with the hot electron injection in FGMOS [92,93]. They started physically modelling both effects following the design of a synapse with long-term weights storage capability [66]. The single transistor synapses simultaneously performed long-term weight storage, computed the product of the input and floating gate value and updated the weight value according to a Hebbian or a backpropagation learning rule, [66,67,69,70,74,92]. The charge in the floating gate was being reduced by using hot electron injection with high selectivity for a particular synapse. It was being increased by using electron tunnelling. As the FGMOS was working in subthreshold and the synapse was very small, this allowed the implementation of small learning systems with a low power consumption. Unlike conventional EEPROMs, reading and writing were now possible simultaneously. Moreover, they also introduced the autozeroing floating gate amplifier (AFGA) as an example of adaptive circuit based on the FGMOS technology, which can learn continuously [68,71,78,81,92].

Other authors devoted their efforts to the design of analog blocks that take advantage of the FGMOS capability to perform the sum operation at the gate. Some examples of these blocks are multipliers, rectifiers, amplifiers and so on [94–103]. Notable is the work of Ramírez–Angulo in low voltage and low power analog design with FGMOS. The transistor also found a niche of application in the fuzzy logic area [104–106]. Ultraviolet light was still used to clean the floating gate.

Lande and Berg's group used the light with a different purpose. They established an ultraviolet conductance between the gate and the drain in the transistor. To achieve this, the whole chip was covered with a second metal layer, except for a hole in the boundaries between the gate and the drain [83–85]. During light exposure supplies were interchanged and certain conditions were set for currents and transconductances, in such a way that the equilibrium point was $V_{DD}/2$ for inputs and outputs when the

MOS floating gates were at the maximum power supply. This initial condition was equivalent to a shift in the threshold voltage values, which allowed the design of circuits working at power supply voltages as low as 0.5 V [107–116].

1.5 Structure of the book

The aim of this book is to present the FGMOS transistor as a powerful device from the circuit design point of view. The inputs in the transistor, significantly more numerous compared with the normal MOS device, appear as extra degrees of freedom which the designer has to play with in order to get a desired functionality. Hence, by establishing the right relationships between them, it is possible to achieve design tradeoffs that are not viable with conventional MOS devices, especially in a context in which power consumption and supply voltage are the main design constraints.

The designs and methodologies in this book are not based on exploiting the physical properties of the device, but just the mathematical ones. Therefore, the reader will not find circuits here whose functionality is based on modifying the charge at the floating gates by using, for example, tunnelling, hot electron injection or UV light. If interested, the reader can find out more about such circuits in the references mentioned in the previous section. All the circuits presented here can be realised in any double poly CMOS process and no extra post-processing is needed.

The book will also show how, with no pretensions to being a panacea, the combination of the FGMOS transistor with LV/LP design techniques provides a good circuit design solution, mostly in situations when the limits of technologies have been pushed far below the values recommended by the manufacturer. In this way, circuits that conventional MOS transistors would render dysfunctional will work when some of these transistors are replaced by FGMOS devices.

The book is mostly about analog design for two reasons: (a) There is more room for creativity in analog circuit design when extra degrees of freedom are available and (b) it is in this context that the device provides greatest advantages. However, a chapter on digital design has been included, because the device also turns out to be ideal for the design of a threshold gate digital block. This can greatly simplify the implementation of digital functions thereby reducing power.

The content is organised as follows:

Chapter 2: 'The Floating Gate MOS transistor: FGMOS'. This chapter introduces the device, starting from the physical structure, followed by some basic device modelling. Subsequently, practical problems that designers traditionally found when starting to work with the device are outlined, together with solutions for these problems.

Chapter 3: 'FGMOS-Circuit applications and design techniques'. This chapter explains the advantages of using an FGMOS device over a standard MOS device in circuits which require low power and low voltage operation. Several simple circuits are presented in order to illustrate the device functionality, explaining the design tradeoffs together with examples.

Chapter 4: 'Low power analog continuous-time filtering based on the FGMOS in the strong inversion ohmic region'. This is the first of four chapters that describe how to exploit the mathematical capabilities of FGMOS devices in more complex circuits, taking as example the design of continuous-time analog filters. In this chapter, the performance of the filter under the low voltage and low power constraints is mainly obtained due to the functionality of FGMOS devices biased in the strong inversion ohmic region. The design methodology is discussed in this, as in the other three chapters, together with a thorough analysis of second-order effects and then design tradeoffs when using the transistor.

Chapter 5: 'Low power analog continuous-time filtering based on the FGMOS in the strong inversion saturation region'. This chapter is quite similar in structure to the previous one, the difference between the two of them being that the topologies presented here achieve low power and low voltage operation by exploiting the properties of FGMOS devices biased in the strong inversion saturation region. Two different designs will be presented, one of them intended for very low frequency and the other one for intermediate frequency applications. Again, the advantages and disadvantages of using the device will be extensively discussed in both. This chapter also shows with an example what happens in a particular design when low quality metal/poly capacitors are used in the device, for those cases in which neither double poly, nor high quality metal/insulator/metal capacitors are available in the technology.

Chapter 6: 'Low power analog continuous-time G_m-C filtering using the FGMOS in the weak inversion region'. This chapter and the next are devoted to exploring the potential of the FGMOS transistor operating in weak inversion for the design of analog low voltage micropower continuous-time filters. This chapter describes a linearised G_m-C topology in which the linearisation is achieved by means of two 'floating gate' blocks. The chapter begins with a reformulation of the translinear principle (TP), which is an ideal technique for implementing nonlinear functions using currents in FGMOS. The advantages of an FGMOS implementation compared with those of a normal MOS implementation in a low voltage context are then discussed. The linearised G_m-C filter is described afterwards followed by an analysis of second-order effects.

Chapter 7: 'Low power log-domain filtering based on the FGMOS transistor'. This chapter will explain how to design a log domain filter using FGMOS transistors. It starts with a review of the basic log domain principle, followed by the description of basic FGMOS blocks that implement the required functions. Again, the advantages and disadvantages of using FGMOS devices in these kinds of topologies will be discussed, together with a qualitative and quantitative analysis of second-order effects.

Chapter 8: 'Low power digital design based on the FGMOS threshold gate'. This chapter mainly focuses on the discussion of an efficient form of implementing digital circuits by using the threshold gate concept. These threshold gates will be implemented with FGMOS transistors. It will be shown how to successfully carry out the aforementioned digital technique of redefining the problem, combined with the utilisation of the FGMOS device, in order to reduce power. This is illustrated with several types of digital circuits grouped in combinational and sequential blocks. All these examples will show how the figure of merit power × delay can be improved with respect to the values obtained with traditional approaches. This chapter is not as incremental as the previous ones, so a previous theoretical background in digital design is recommended.

Chapter 9: 'Summary and conclusions'. This chapter brings together the main points of each chapter and revisits the subject relevance.

Since the number of degrees of freedom that need to be handled in all these designs is larger than in normal approaches, the mathematics in all these chapters utilise a considerably large number of variables. Because of this a notation section can be found at the end of each chapter for the reader to refer to as needed.

Chapter 2

The Floating Gate MOS transistor (FGMOS)

2.1 Introduction

This chapter introduces the Floating Gate MOS transistor (FGMOS). The properties of this device are described and a simple model for hand analysis is presented. Throughout the chapter, the FGMOS transistor is compared with a standard MOS device, and the main advantages and disadvantages of using an FGMOS instead of an MOS transistor are drawn. Also, some of the common initial problems a designer faces when trying to use the FGMOS are outlined and solutions for them are provided. Hence, in this chapter, the reader will find out how to perform accurate simulations of circuits containing FGMOS devices without having to model them. Also, the frequently asked question 'how to effectively eliminate the charge accumulated at the floating gate during the fabrication process' is reviewed and answered in one of the sections. Finally, a quantitative analysis of the minimum extra area required by an FGMOS when compared with an MOS with the same channel size is presented in the final sections.

2.2 The Floating Gate MOS (FGMOS) device

2.2.1 Introducing the device

An FGMOS can be fabricated by electrically isolating the gate of a standard MOS transistor, so that there are no resistive connections to its gate. A number of secondary gates or inputs are then deposited above the floating gate (FG) and are electrically isolated from it. These inputs are only capacitively connected to the FG, since the FG is completely surrounded by highly resistive material. So, in terms of its DC operating point, the FG is a floating node.

A possible layout of a three-input n-channel FGMOS transistor together with three cross-sectional views are shown in Fig. 2.1[3], and will be used to illustrate

---

[3] Fig. 2.1 is not drawn to scale.

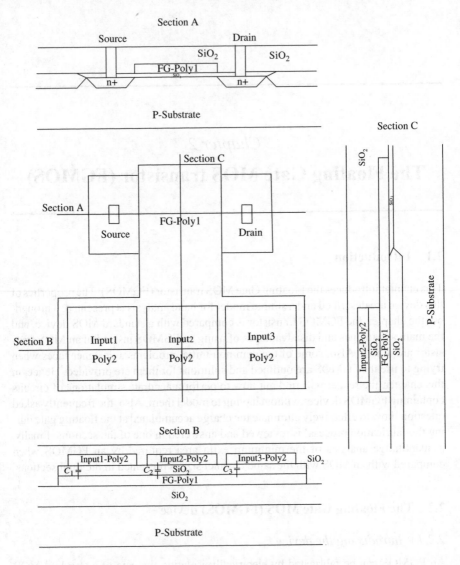

Figure 2.1   Possible layout of a 3-input n-channel FGMOS transistor (middle) and its cross-sectional views (A, B and C)

important features of FGMOS transistors as well as to show how they can be fabricated in existing MOS technologies. The equivalent schematic for an N-input n-channel FGMOS transistor is given in Fig. 2.2 whilst Fig. 2.3 shows the schematic symbol that will be used throughout this book.

It can be seen in Fig. 2.1 how the FG, fabricated using the gate electrode (poly1) layer, extends outside the active area of the MOS transistor. The FG is surrounded by two SiO$_2$ insulator layers and thus electrically isolated from the rest of the device.

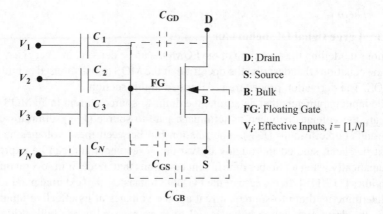

Figure 2.2   Equivalent schematic for an n-channel N-input FGMOS transistor

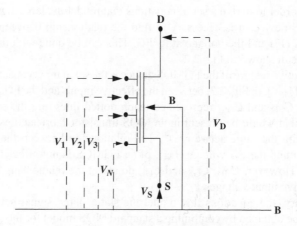

Figure 2.3   Symbol and voltage definitions for an n-channel N-input FGMOS

The device inputs are placed on top of the upper SiO$_2$ insulating layer and are fabricated using another conducting layer, preferably a second layer of polysilicon (poly2). The sizes of the input electrodes (Fig. 2.1) determine the values of the capacitors that connect the FGMOS inputs with the FG and they can be varied according to the designer's needs.

The values of the input capacitances are given by

$$C_i = \left(\frac{\varepsilon_{SiO_2}}{t_{SiO_2}} \right) A_i \tag{2.1}$$

where ε_{SiO_2} is the permittivity of the SiO$_2$, t_{SiO_2} is the thickness of the SiO$_2$ between the FG and the effective inputs and A_i is the area of each input capacitor plate.

2.2.2   Theory

2.2.2.1   Large signal DC behaviour

Equations modelling the operation of the FGMOS can be derived, in a very easy way, from the equations that describe the operation of the MOS transistor used to build the FGMOS. The derivation procedure is described in this section.

The input parameters that determine the drain-to-source current in an MOS transistor are the voltages between its terminals: gate-to-source (V_{GS}), drain-to-source (V_{DS}) and source-to-bulk (V_{SB}). The relationship between these voltages and the current has been studied thoroughly by various researchers and can be expressed mathematically using a number of different equations that model it in several operating regimes [117,118]. In the case of the FGMOS transistor, the 'external parameters' that determine its drain-to-source current are the voltages at its effective inputs, as well as the drain and source voltages, all of them referred to the bulk (Fig. 2.3). If it is possible to determine the voltage at the FG of an FGMOS device, it is then possible to express its drain to source current using standard MOS transistor models. Therefore, in order to derive a set of equations that model the large signal operation of an FGMOS device, it is necessary to find the relationship between its effective input voltages (V_i) and the voltage at its FG. This can be done with the help of the equivalent circuit shown in Fig. 2.2.

The distinctive feature of the FGMOS device is the set of input capacitors, denoted C_i where $i = [1, N]$, in Fig. 2.2, between the effective inputs and the FG. The parasitic capacitances, C_{GD} and C_{GS}, represented using dotted lines, are the same parasitic capacitances that would be present in an MOS transistor fabricated using the same technology with the same active area[4]. The relationship between the DC drain to source current and the FG voltage, V_{FG}, of an FGMOS is not affected by parasitic capacitances. However, C_{GD}, C_{GS} and C_{GB} do affect the relationship between V_{FG} and the effective input voltages V_i.

In summary, the equations that model the static, large signal behaviour of an FGMOS can be obtained by combining a standard MOS model for the same technology with the equation that relates V_{FG} to V_i, V_D, V_S, C_i, C_{GD}, C_{GS} and C_{GB}. This equation can be obtained by applying the charge conservation law to the floating node (FG) shown in Fig. 2.2. If there is an infinite resistance between the FG and all the surrounding layers, there will be no leakage current between them, and so, the FG will be perfectly isolated. Under this assumption the voltage at the FG is given by

$$V_{FG} = \sum_{i=1}^{N} \frac{C_i}{C_T} V_i + \frac{C_{GS}}{C_T} V_S + \frac{C_{GD}}{C_T} V_D + \frac{Q_{FG}}{C_T}$$

$$= \sum_{i=1}^{N} \frac{C_i}{C_T} V_{iS} + \frac{C_{GD}}{C_T} V_{DS} + \frac{C_{GB}}{C_T} V_{BS} + \frac{Q_{FG}}{C_T} + V_S \qquad (2.2)$$

---

[4] C_{GB} is not the same, which will be explained later in this chapter.

where N is the number of effective inputs. The term Q_{FG} refers to a certain amount of charge that has been trapped in the FG during fabrication. As this term is constant, it can be interpreted as a voltage offset at the FG, or alternatively, as an offset in the threshold voltage of the device. The term C_T refers to the total capacitance seen by the FG and is given by

$$C_T = C_{GD} + C_{GS} + C_{GB} + \sum_{i=1}^{N} C_i \tag{2.3}$$

The equations modelling the large signal behaviour of the FGMOS can now be obtained by replacing V_{GS} in the equations describing the large signal behaviour of the MOS transistor, with the expression describing the voltage between the FG and source which can be obtained by referring V_{FG} to the source terminal rather than the bulk. This procedure is illustrated in Table 2.1. The left-hand column shows typical model equations for hand calculations for an n-channel MOS transistor, and the right-hand column shows the equivalent expressions for the FGMOS. W and L are the effective width and length of the transistor. The undefined parameters have their usual meaning [119].

2.2.2.2 Small signal equations

An N-input FGMOS device has $N - 1$ more terminals than an MOS transistor, and therefore, $N+2$ small signal parameters can be defined: N effective input transconductances, g_{mi}, for $i = [1, N]$, an output conductance g_{dsF} and a bulk transconductance g_{mbF}. If g_m, g_{ds} and g_{mb} are the gate transconductance, output conductance and bulk transconductance, respectively, of an MOS transistor having identical channel size, current and V_{DS} as the FGMOS, the FGMOS small signal parameters are then related to the MOS parameters in the following way[5]:

$$g_{mi} = \frac{C_i}{C_T} g_m \quad \text{for} \quad i = [1, N] \tag{2.10}$$

$$g_{dsF} = g_{ds} + \frac{C_{GD}}{C_T} g_m \tag{2.11}$$

$$g_{mbF} = g_{mb} + \frac{C_{GB}}{C_T} g_m \tag{2.12}$$

These equations show two drawbacks of the FGMOS compared with the MOS transistor: the **reduction of the input transconductance** and the **reduction of the output resistance**. In order to illustrate this, let us take as an example the common source amplifier shown in Fig. 2.4(a). Its gain (G) and output resistance (R_{out}) are

$$G = g_m^{(3)} \left(g_{ds}^{(2)} + g_{ds}^{(3)} \right)^{-1} \quad R_{out} = \left(g_{ds}^{(2)} + g_{ds}^{(3)} \right)^{-1} \tag{2.13}$$

---

[5] The term g_{mi} will be employed to refer to the effective transconductance only in this chapter. For the sake of clarity in the notation, its value, $C_i/C_T \cdot g_m$ will be used instead in subsequent chapters.

Table 2.1   MOS versus FGMOS model equations

MOS	FGMOS

Weak inversion saturation when $V_{BS} = 0$

$I_D = I_s e^{(V_{GS}/nU_T)}$ (2.4)

$$I_D = I_s e^{\left(\sum_{i=1}^{N} \frac{C_i}{C_T} \frac{V_{iS}}{nU_T}\right)} e^{\frac{C_{GD}}{C_T} \frac{V_{DS}}{nU_T}} e^{\frac{Q_{FG}}{nU_T C_T}}$$ (2.5)

for $V_{GS} < V_T$, $V_{BS} = 0$

$V_{DS} > 4U_T$

for $\sum_{i=1}^{N} \frac{C_i}{C_T} V_{iS} + \frac{C_{GD}}{C_T} V_{DS} + \frac{Q_{FG}}{C_T} < V_T$

$V_{DS} > 4U_T$, $V_{BS} = 0$

Strong inversion ohmic

$$I_D = \mu_0 C_{ox} \frac{W}{L}$$
$$\times \left[(V_{GS} - V_T) - \frac{V_{DS}}{2} \right] V_{DS}$$
(2.6)

$$I_D = \mu_0 C_{ox} \frac{W}{L} \left\{ \left[\left(\sum_{i=1}^{N} \frac{C_i}{C_T} V_{iS} \right) - \left(V_T - \frac{C_{GB}}{C_T} V_{BS} - \frac{Q_{FG}}{C_T} \right) \right] - \left(\frac{1}{2} - \frac{C_{GD}}{C_T} \right) V_{DS} \right\}$$
(2.7)

for $0 < V_{DS} \leq (V_{GS} - V_T)$

$V_{GS} > V_T$

for $0 < V_{DS} \leq \left(\sum_{i=1}^{N} \frac{C_i}{C_T} V_{iS} + \frac{C_{GD}}{C_T} V_{DS} \right.$

$\left. + \frac{C_{GB}}{C_T} V_{BS} + \frac{Q_{FG}}{C_T} - V_T \right)$

$V_{GS} > V_T$

Strong inversion saturation

$$I_D = \frac{\mu_0 C_{ox}}{2} \frac{W}{L} (V_{GS} - V_T)^2$$
$$= \frac{\beta}{2} (V_{GS} - V_T)^2$$ (2.8)

for $0 < (V_{GS} - V_T) \leq V_{DS}$

$V_{GS} > V_T$

$$I_D = \frac{\mu_0 C_{ox}}{2} \frac{W}{L} \left(\sum_{i=1}^{N} \frac{C_i}{C_T} V_{iS} + \frac{C_{GD}}{C_T} V_{DS} \right.$$
$$\left. + \frac{C_{GB}}{C_T} V_{BS} + \frac{Q_{FG}}{C_T} - V_T \right)^2$$
$$= \frac{\beta}{2} \left(\sum_{i=1}^{N} \frac{C_i}{C_T} V_{iS} + \frac{C_{GD}}{C_T} V_{DS} + \frac{C_{GB}}{C_T} V_{BS} + \frac{Q_{FG}}{C_T} - V_T \right)^2$$

$$0 < \left(\sum_{i=1}^{N} \frac{C_i}{C_T} V_{iS} + \frac{C_{GD}}{C_T} V_{DS} + \frac{C_{GB}}{C_T} V_{BS} \right.$$ (2.9)
$$\left. + \frac{Q_{FG}}{C_T} - V_T \right) \leq V_{DS}$$

$V_{GS} > V_T$

where the superscript (*i*) refers to the transistor named Mi. If the input device is substituted by an FGMOS, the new values for the gain and output resistance will be

$$G_{FG} = w_1 g_m^{(3)} \left[g_{ds}^{(2)} + g_{ds}^{(3)} + \frac{C_{GD}}{C_T} g_m^{(3)} \right]^{-1}$$

$$R_{outFG} = \left(g_{ds}^{(2)} + g_{ds}^{(3)} + \frac{C_{GD}}{C_T} g_m^{(3)} \right)^{-1}$$ (2.14)

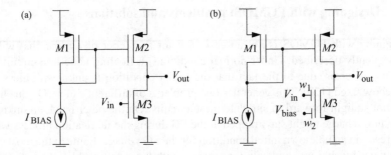

Figure 2.4 *Common source amplifier: (a) With an MOS transistor at the input. (b) With an FGMOS transistor at the input*

where w_i are the equivalent input weights (C_i/C_T) for the FGMOS transistor (Fig. 2.4(b)). Hence if, for example, the circuit in Fig. 2.4(a) is designed with a gain of 100 and output resistance of 1 MΩ and the same circuit is designed by replacing the input MOS transistor by an FGMOS device with two identical inputs $(w_1 = w_2)$, $C_{GD}/C_1 = 0.005$, $C_{GS}/C_1 = 0.05$ and $C_{GB}/C_1 = 0.1$ $(w_1 = w_2 = C_1/(2C_1 + C_{GD} + C_{GS} + C_{GB}) = 0.464)$, the new gain would then be $G_{FG} = 38$ and the output resistance $R_{outFG} = 812$ kΩ. However, benefits of FGMOS often overweight these disadvantages.

2.2.2.3   Noise and dynamic range

The sources of noise in the FGMOS are the same as in the MOS transistor since the noise is generated by the movement of carriers in the channel and the carriers are the same in both devices. However, the equivalent noise at the input of the FGMOS will be different, since the signal at the FG is attenuated, or, what is equivalent, it is amplified at the effective input. Hence, the equivalent power spectral density of the noise at an effective input will be

$$\left.\frac{\overline{v_{in}^2}}{\Delta f}\right|_{\text{effective input}} = \left(\frac{1}{w_i}\right)^2 \cdot \left.\frac{\overline{v_{in}^2}}{\Delta f}\right|_{FG} \tag{2.15}$$

where w_i is the capacitive weight corresponding to the input i and $(\overline{v_{in}^2}/\Delta f)|_{FG}$ is the noise power spectral density at the FG. Equation (2.15) shows how the equivalent noise increases at the effective input of an FGMOS device, since the factor $(1/w_i)$ is greater than 1. However, in the worst case scenario, the DR will not be affected because the maximum allowable input signal for a certain level of distortion will also be increased by the same factor $(1/w_i)$. The reason for this is the following: if the input capacitances are assumed to be linear, the signal seen by the FG is actually the signal at the effective input of the FGMOS attenuated by a factor w_i.

In general though, the DR will even improve since, for example, from the system level point of view, the use of the FGMOS will greatly simplify the circuit structure. This will consequently reduce the distortion as well as the noise levels since the number of devices and hence the number of noise contributors will be smaller.

2.3   Designing with FGMOS: problems and solutions

Designing with FGMOS is not an easy task for a number of reasons that will be subsequently described. Under normal conditions, a floating node in a circuit represents an 'error' due to the fact that the initial condition is unknown unless it is somehow fixed [120]. This generates two problems of different nature. On one hand, it is not straight forward to simulate these circuits. On the other hand, an unknown amount of charge might stay trapped at the FG during the fabrication process which will result in an unknown initial condition for the FG voltage. Some of the most commonly used techniques and solutions to deal with these two problems are discussed in this section. Their advantages and disadvantages are illustrated with examples.

2.3.1   Simulation

The first problem any designer finds when dealing with FGMOS is the lack of simulation models. Because of this, the simulators do not understand the FG and report it as a convergence error [120]. Several solutions to this problem have been offered in literature [121–124]. The most popular ones are briefly described here putting a particular emphasis on the technique that offers the most reliable prediction of the FGMOS operation [124].

The first solution to the convergence problem was offered by Ramírez-Angulo *et al.* in [121]. It is based on the simulation model shown in Fig. 2.5. This model overcomes the simulation problem by connecting a very high value resistor (R_G) between the FG node and a set of voltage controlled voltage sources (VCVS) that model the addition at the FG as described by eq. (2.2).

The gains (a_i, a_o and a'_o) of the VCVSs are the ratios between the corresponding input capacitances and the total capacitance seen by the FG. The simulator calculates

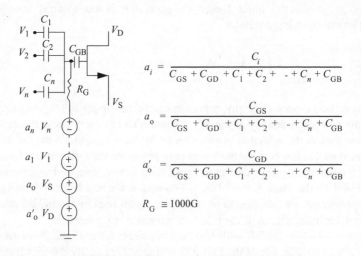

$$a_i = \frac{C_i}{C_{GS} + C_{GD} + C_1 + C_2 + \ \cdot \ + C_n + C_{GB}}$$

$$a_o = \frac{C_{GS}}{C_{GS} + C_{GD} + C_1 + C_2 + \ \cdot \cdot \ + C_n + C_{GB}}$$

$$a'_o = \frac{C_{GD}}{C_{GS} + C_{GD} + C_1 + C_2 + \ \cdot \ + C_n + C_{GB}}$$

$$R_G \cong 1000G$$

Figure 2.5   Ramírez-Angulo's model for simulation of the FGMOS

the DC operating point by annulling the capacitances. This creates an open circuit at the gate. As no current is flowing through the resistance R_G, the FG DC operating point is given by the sum of the VCVS voltages. In AC analysis, the VCVSs do not have any influence on the simulation because the resistance connected to the gate is very large and thus behaves as an open circuit even for very low frequencies.

Although very commonly used, this approach has several important limitations. On one hand, the parasitic capacitances have to be known from the beginning. The only way to obtain their values is to, prior to application of the proposed method, carry out a set of simulations with an estimated operating point at the FG. This is not completely correct since the voltage at the FG does depend on these parasitic capacitances. Besides, once their values are obtained this method assumes that they are constant for the whole operating range, which is not always true. In fact the deviations from these values can be quite significant if the transistor has to operate in more than one region. Also, the final values of the parasitics connected to the FG increase once the circuit has been laid out and this can be difficult to model in the schematics [125]. All these facts increase the risk of failure in certain topologies, and is critical in those analog cells that have to work with reduced voltage margins, such as, for example, the low voltage circuits described along this book. Circuits operating in the subthreshold region are also very much affected by the variation of the parasitic capacitances since this variation will have an exponential effect.

A very important consequence of an incorrect estimation of C_{GD} is the miscalculation of the output resistance (see eq. (2.11)), which can cause a number of undesired effects such as unexpected losses in a bandpass filter, large variations in a quality factor and cutoff frequency, instability and so on.

All these problems could be minimised by oversizing the transistors in such a way that the parasitic capacitance which varies the most is much smaller than the minimum input capacitance (how much smaller would depend on the value of the expected variation and how it could affect the circuit performance). However, it is not easy to know a priori how much bigger the input capacitors need to be. Previous assumptions need to be made that are often not trivial and could even give rise to undesired consequences such as instability in circuits with feedback. Besides, oversizing the input capacitors might be unnecessary and therefore a waste in terms of area.

Yin *et al.* suggest a different model to simulate FGMOS transistors [122]. The model is based on connecting resistors in parallel with the input capacitors as shown in Fig. 2.6. The equation for the operating point is the same one as in Ramírez-Angulo's approach. The problems of this model are also the same.

Another technique proposed in literature [123] is based on iterative simulations which are directly performed by the simulator with the help of a program that has to be previously implemented. This can be done using the SKILL language functions in Cadence [123,125]. This technique is a more accurate, but still not exact enough, version of the method in [121].

A reliable technique to simulate FGMOS transistors is illustrated in Fig. 2.7 [124]. It is based on the use of an initial transient analysis (ITA). At the beginning of this analysis supply voltages and circuit inputs are set to 0 V. Under these conditions,

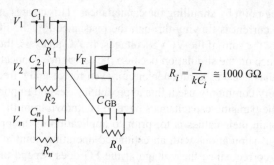

Figure 2.6   Model in [122] for the FGMOS

$$R_i = \frac{1}{kC_i} \cong 1000 \text{ G}\Omega$$

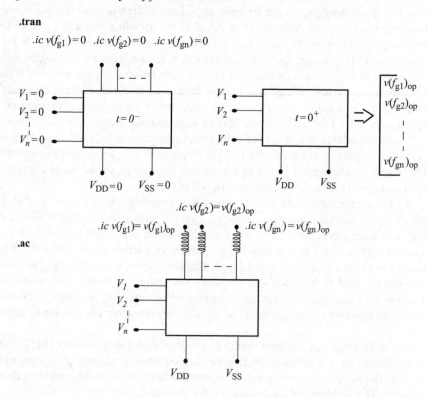

Figure 2.7   Simulation technique based on an initial transient analysis (ITA)

it can be guaranteed that the initial voltages at the FGs are also zero, unless that, as it happens in some technological processes or under certain conditions, some residual charge (Q_{FG}) remains trapped at the FG after fabrication. If this is the case, the initial condition at the FGs would be Q_{FG}/C_T. In any case, this is an unpredictable term common to all models and will be discussed in more detail in the next section. The transient analysis starts with these initial conditions. If, for example, HSpice is

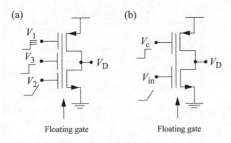

Figure 2.8   *(a) Circuit used to illustrate the simulation technique in Fig. 2.7*
           (b) Circuit used to compare the techniques in Figs. 2.5 and 2.7

used, the initial conditions can be imposed with the command .ic [120]. Subsequently, the supply voltages are set to their final values and the inputs evolve normally. Small signal simulations can be performed extracting the operating point $(v(f_{gi})_{op}$ for $i =$ $[1, n])$ from the transient analysis and forcing it to the FGs with the help of very high value inductances. The inductances behave as wires in DC and fix the operating point. Besides, if they are big enough, their effect is negligible for any other operating frequency. In the case of Hspice, for example, after completing a transient simulation, the operating point can be found in the file input_file_name.ic (where input_file_name is the name of the input file). In order to force the initial conditions the command .nodeset at the beginning of the file needs to be changed to .ic.

An example of how to use this simulation technique is provided below. The netlist describes the FGMOS inverter shown in Fig.2.8(a). First of all, the subcircuits for generic n-channel and p-channel transistors are defined. The external nodes for these subcircuits are the FGs, drains, effective inputs and sources. The parameters are the values of the input capacitances and the dimensions of transistors. The reason for adding the FG to the list of external nodes, despite the fact that it has to be completely isolated, is that it gives the freedom to change the number of capacitive inputs without having to define a new device. The simulation shows how to perform a parametric DC analysis. As it was previously explained, voltage supplies and inputs are initially set to 0 V and a zero initial condition is forced at the FG with the option .ic. When the circuit is 'powered up' [vdd vdd 0 pulse(0 2 0.01 0.01 1 2 4)] , the inputs that are going to be constant in the steady state change quickly [v1 v1 0 pulse(0 a 0.01 0.01 1 2 4), v3 v3 0 pulse(0 2 0.01 0.01 1 2 4)], whereas the input which is supposed to be swept in DC rises very slowly [v2 v2 0 pulse (0 2 0.01 1 1 2 4)]. In subsequent simulations the final value of the DC bias ('a' parameter in v1) is set to different values thus performing the parametric variations. The input–output characteristics obtained with these simulations are shown in Fig. 2.9.

Netlist for the circuit in Fig. 2.8(a)

```
Inverter

.include 'c:\simulation\modelos\NC\Typical\NC.sp'
```

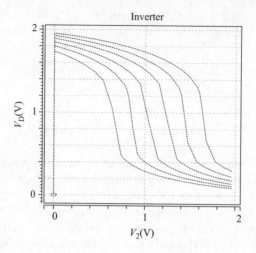

Figure 2.9   Input–output characteristic for the FGMOS inverter in Fig. 2.8(a)

```
.include 'c:\simulation\modelos\PC\Typical\PC.sp'

.subckt transistorn vg vd v1 v2 v3 vs w=20u l=20u
    c1=210f c2=210f c3=210f

m1 vd vg vs 0 nc w='w' l='l'

c1 v1 vg c1

c2 v2 vg c2

c3 v3 vg c3

.ic vg=0

.ends

.subckt transistorp vg vd v1 v2 v3 vs w=20u l=20u
    c1=210f c2=210f c3=210f

m1 vd vg vs 0 pc w='w' l='l'
```

```
c1 v1 vg c1

c2 v2 vg c2

c3 v3 vg c3

.ic vg=0


.ends

x1 vg vd v1 v2 v3 0 transistorn w=20u l=20u c1=210f
    c2=210f c3=210f

x2 vg vd v1 v2 v3 vdd transistorp w=20u l=20u c1=210f
    c2=210f c3=210f


v1 v1 0 pulse(0 a 0.01 0.01 1 2 4)

v2 v2 0 pulse(0 2 0.02 1 1 2 4)

v3 v3 0 pulse(0 2 0.01 0.01 1 2 4)

vdd vdd 0 0 pulse(0 2 0.01 0.01 1 2 4)


.tran 0.01 1 sweep a 0 10 2

.op t=1

.options post


.end
```

The ITA technique is even easier to apply if the Spectre simulator is used in Cadence to design the circuit [125]. In this case the input file does not need to be modified to perform an AC simulation. The latter can be realised just by choosing the simulation options adequately in a transient simulation, followed by an AC analysis. In the initial transient analysis all the sources will be automatically set to zero by the simulator. They will then rise to their final value between the 'starttime' and the time zero. At the final time the operating point will be saved in a file. The AC analysis will then be run using this initial operating point from the file.

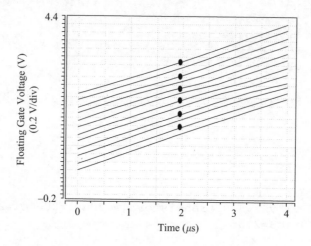

Figure 2.10   Voltage at the FG of the CMOS inverter in Fig. 2.8(b) obtained with the VCVSs model and with the ITA technique

Figure 2.10 compares the ITA technique with the technique in [121] (see Fig. 2.5) using as an example the input–output characteristic of the inverter in Fig. 2.8(b). The evolution of the FG voltage is shown with one of the inputs being continuously swept and the other one changing parametrically. The dashed lines marked with spots are the results obtained using the ITA technique. The other lines are obtained using Ramirez-Angulo's method. There is a variation of around 0.2 V that demonstrates how the latter is not very reliable and depends very much on the designer's ability to determine the parasitic capacitances accurately.

Figure 2.11 shows the output resistance of an FGMOS transistor obtained with a transient analysis when either (a) the VCVSs technique or (b) the ITA technique is used. It can be seen how the VCVSs technique generates a value around five times smaller. This is due to a bad estimation of the C_{GD} value.

2.3.2   Charge accumulation

Until recently FGMOS transistors had only been used in digital electronics EEPROM devices [126] and despite the important role FGMOS devices could play in analog circuits, designers have been reluctant to work with them. The main reason for this is the uncertain amount of charge (Q_{FG}) that might stay trapped at the FG during the fabrication process causing variations of the threshold voltage. Q_{FG} has different implications on different circuits; however, the design will often not work unless the charge is removed. Reported solutions to this problem include cleaning with ultraviolet (UV) light; the use of tunnel effect and hot electron injection; and forcing an initial condition with a switch [42,83,84,92,127].

UV cleaning refers to exposing the FG to UV light. It is well known that when the surface of any semiconductor is exposed to light or any other electromagnetic radiation, part of this radiation is reflected, part of it is absorbed and the rest is

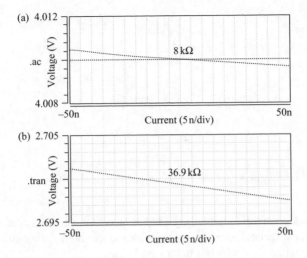

Figure 2.11   Comparison between the output resistances of an FGMOS obtained using (a) VCVSs technique (b) ITA technique

transmitted [83,84]. The number of absorbed photons is proportional to the total number of photons, and therefore to the light intensity. It also depends on the type of semiconductor, the wavelength of the photons and the applied electric field. When the FG is illuminated with UV light, the electrons which are trapped at the gate are able to travel through the potential barrier in the oxide/silicon interface. Hence, when the light illuminates the gate/source or gate/drain region of the MOS transistor, the excited electrons will gain enough energy to traverse the gate oxide barrier. One of the main drawbacks of this technique is in its incompatibility with those CMOS technologies whose passivation layers are not transparent to UV light. In such cases, in order to enable light transmission, the passivation has to be eliminated in the fabrication process. Unfortunately, this can affect the lifetime of the chip.

Fowler–Nordheim tunnelling is the process whereby electrons tunnel through a barrier in the presence of a high electric field. The main drawback of the tunnelling technique is that, in order to generate Fowler–Nordheim currents, a very high electric field has to be generated in the region between the drain and the gate. The magnitude of the electric field is inversely proportional to the thickness of the oxide underneath the gate and directly proportional to the voltage across it. Therefore, unless the oxide is very thin, the voltage needs to be extremely high. Hence, for example, for an oxide thickness in the order of 10 nm, the voltage required is around 15 V. Besides, extra circuitry is required in order to control the intensity and duration of the Fowler–Nordheim currents and thus the exact amount of charge at the FG.

Hot electron injection can be generated with smaller voltages. However, the injection grows with the current in the channel. Hence, the currents have to be very small in order to avoid huge variations. Again, the process has to be controlled internally with additional circuitry that will be different for each system.

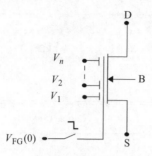

Figure 2.12   Technique to set the initial conditions at the FG

A drawback common to tunnelling and hot electron injection is that they are not modelled in most technologies. Hence, in order to use them efficiently they would have to be modelled first, which is not a simple task.

In any case, although these techniques (tunnelling and hot electron injection) are not recommended when the intention is to erase the initial fabrication charge (due to the complexity they add to the overall circuit topology), they have proven to be a very interesting alternative when adaptability or programmability is required [73–81,127,129]. Nevertheless, these kinds of applications are far from the context of this book and therefore will not be covered in more detail.

An alternative technique to control the charge at the FG was presented in [62]. This technique uses a switch to electrically set the initial operating point at the FG by short-circuiting it to a certain voltage ($V_{FG}(0)$) for a given combination of inputs (Fig. 2.12). This solution deleteriously influences the device operation. For example, if the switch is introduced in such a way as to enable discharging of the FG, then subsequently, during normal FGMOS operation, this switch will provide a large but finite resistance path from the FG to the substrate of the switch. So, the gate will no longer be floating and its DC voltage will drift towards the switch substrate voltage. This technique has been successfully used in digital and sampled-data applications, but it is not suitable for general purpose continuous-time analog circuits.

The best results so far have been achieved with the technique in [130]. This technique reduces Q_{FG} down to values very close to zero. It consists of adding a series of contacts from poly1 to the top metal layer. The top metal layer is the one that is deposited and etched last. The contacts do not connect the FG to any other part of the circuit, so functionally, they do not alter the device. However, the added contacts resolve the trapped charge problem. A possible explanation for this is that the fabrication stages prior to the deposition of the top metal layer lead to an accumulation of trapped charge on the FGs. Before etching takes place all parts of the chip in contact with the top metal layer are connected together. Since all the FGs are connected to the latter via the poly1 to metal contacts, the charge trapped at each gate flows to other parts of the die. For example, if the substrate is connected to the top metal layer then the FGs might discharge to the substrate via these contacts. The key is that during one of the final stages of fabrication, the FGs are not floating. Then, when the top

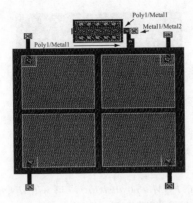

Figure 2.13   Four-input FGMOS transistor layout with gate-to-metal contacts to eliminate the charge in a two-metal layer process

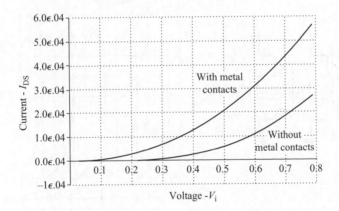

Figure 2.14   Comparison of measured results for two identical floating gate transistors, one with discharging contacts and another without them

metal is etched away, their floating nature is restored. The layout in Fig. 2.13 shows how this can be done.

The poly1 gate of the FGMOS is connected to the extra region of poly1 via a metal1 bridge, which is only required here because the two poly1 regions are disconnected. The poly1 to metal1 contact in conjunction with the metal1 to metal2 contact provides the connection from the FG to the top metal layer in this technology, i.e. metal2.

As an example of the efficiency of this technique, Fig. 2.14 shows the experimental results for two equally sized 1-input n-channel FGMOS transistors, one with a contact at the FG and the other without it. The transistors were fabricated in a 0.5 μm CMOS technology. The graph shows the drain-to-source current (I_{DS}) versus the voltage at the effective input. The effective threshold voltage is around 200 mV higher in the transistor with no contacts at the FG.

Figure 2.15   Two 2-input FGMOS transistors: one with discharging contacts (right) and another without them (left)

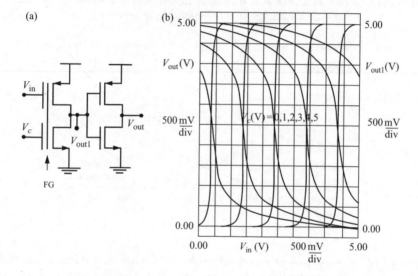

Figure 2.16   (a) Schematic of the circuit used to check the charge accumulation (b) Input–output characteristic of the circuit in (a) – experimental results

Figure 2.15 shows a microphotograph of two 2-input FGMOS transistors; one with contacts at the FG on the right and another one without them on the left.

Another example which is illustrated in Fig. 2.16 shows the performance of a simple 2-input FGMOS inverter followed by an identical inverter designed using conventional MOS transistors. Ignoring the effects of conventional mismatch in the transistors, the switching threshold of both inverters will be the same in the absence of trapped charge. Fig. 2.16(b) shows experimental results obtained when sweeping one of the inputs between 0 and 5 V and varying the other input in steps of one volt. The difference in thresholds is within a few millivolts. These small differences can be attributed to conventional transistor mismatch and they do not change the circuit

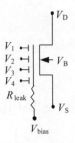

Figure 2.17   Quasi-FGMOS equivalent circuit

performance. The transfer function of the same inverter fabricated without the metal k contact would be saturated for the whole range of input voltages.

All the fabricated circuits presented along this book were designed using this technique in several technologies over a number of runs. In all the cases the experimental results were in agreement with the expected performance. No charge accumulation was detected, whereas circuits fabricated without using this technique did not work at all.

2.3.2.1   Pseudo-FGMOS: when the charge is not a problem

A modified version of the FGMOS in which the initial charge is no longer an issue is shown in Fig. 2.17 [131].

This device, called, pseudo- or quasi-FGMOS, is an FGMOS transistor whose gate is tied to a very large value resistor that weakly connects it to the most positive or most negative voltage value within the circuit (depending on whether it is an n- or a p-channel device). Figure 2.17 represents a 4-input n-channel quasi-FGMOS. For a generic N-input quasi-FGMOS, the voltage at the gate is

$$V_{\mathrm{FG}} = \frac{sR_{\mathrm{leak}}}{1 + sR_{\mathrm{leak}}C'_{\mathrm{T}}} \left(\sum_{i=1}^{N} C_i \cdot V_i + C_{\mathrm{GS}} \cdot V_{\mathrm{S}} + C_{\mathrm{GD}} \cdot V_{\mathrm{D}} \right) \qquad (2.16)$$

where all the voltages have been referred to the bulk and C'_{T} is the total capacitance seen by the gate of the quasi-FGMOS. C'_{T} will be different from C_{T} in the normal FGMOS as in reality R_{leak} will be implemented using active devices. These devices will have parasitic capacitances that contribute to the value of C'_{T}. It follows from eq. (2.16) that the inputs are highpass filtered with a cutoff frequency which is inversely proportional to R_{leak}. Hence, as long as R_{leak} is kept high enough, the gate can be effectively floating for very low frequency values so that the AC operation is unaffected. A couple of techniques to implement R_{leak} were originally presented in [132] and [133]. They are illustrated in Fig. 2.18(a) and Fig. 2.18(b) for an n-channel FGMOS.

Both techniques suffer from the same drawback: the gate voltage must not exceed the rail by more than the cut-in voltage of the p-n body–source junction of the nMOS transistor realising R_{leak}, so that it does not become forward biased. A technique that

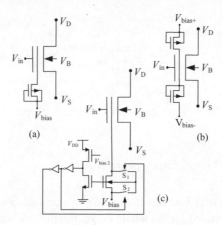

Figure 2.18   (a) Implementation of the quasi-FGMOS proposed in [131] (b) Implementation of the quasi-FGMOS proposed in [132] (c) Wide range quasi-FGMOS

improves the input range is illustrated in Fig. 2.18(c) using an n-channel FGMOS transistor. R_{leak} is implemented using an nMOS transistor with its gate and bulk connected together. The drain/source is connected to a bias voltage V_{bias}, whereas the source/drain is tied to the gate. Two switches (S_1 and S_2) are used to control the voltage at the bulk so it never exceeds the voltage at the drain. In DC the device will operate as a normal quasi-FGMOS. The DC value of the gate voltage will be V_{bias} and the DC component of the signal will be filtered out. In AC the transistor will work as a conventional FGMOS. For negative excursions of the input signal the source of the transistor implementing R_{leak} will be the terminal connected to the quasi-floating gate. In this case, switch S_1 will be on, S_2 will be off and the bulk will be connected to it. For positive excursions of the input signal, the terminal connected to the quasi-floating gate will act as the drain, switch S_2 will be on and switch S_1 will be off. In this way the p-side of the pn junctions will always be connected to the most negative voltages and will never be forward biased. The comparison between the gate voltage and V_{bias} can be performed using a simple common-source amplifier whose threshold can be changed through the bias current. The exact value is not important since the voltage margin before the output resistance of the diode starts dropping is a few hundred millivolts. The output of the latter is converted to digital values by two CMOS inverters that control the switches. For the sake of simplicity S_1, S_2 and the two inverters have been represented as ideal components in Fig. 2.18(c). However, the results shown later on were obtained with real nMOS switches and CMOS inverters.

Figure 2.19 compares the performance of the three versions of quasi-FGMOS transistor. The voltage at the gate changes from 0 to 2 V. The maximum value of the leakage current for the R_{leak} implementation in Fig. 2.18(c) is of the order of 10^{-13} A (Fig. 2.19(a)), whereas this current can reach values above mA for gate voltages below 0.5 V (Fig. 2.19(c)) for the topology in Fig. 2.18(a) and above 1.75 V for the one in Fig. 2.18(b) (Fig. 2.19(b)). Besides, the value of R_{leak} varies over several

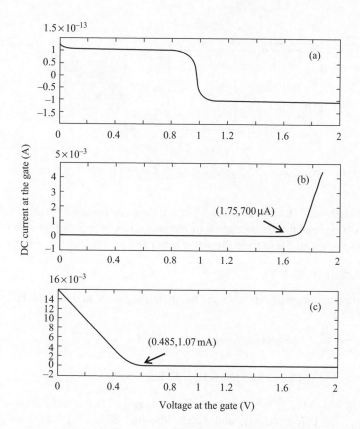

Figure 2.19   *(a) DC current versus voltage at the gate for the implementation in Fig. 2.18(c) (b) DC current versus voltage at the gate for the implementation in Fig. 2.18(b) (c) DC current versus voltage at the gate for the implementation in Fig. 2.18(a)*

orders of magnitude within the input range for the implementations in Fig. 2.18(a) and Fig. 2.18(b), whereas it remains in the order of 10^{13} Ω for the design in Fig. 2.18(c).

Being able to use the whole voltage range is of uttermost importance in analogue designs that have to operate at a low voltage supply. Besides, the DC value of the signal can be shifted through V_{bias} which means that every device can be biased in the most convenient operating region.

2.4   Minimum input capacitance: implications to the total area of the device

One disadvantage of the FGMOS transistor compared with the MOS transistor is in its much larger area due to the added input capacitors. The size of the transistor depends on a number of factors that are discussed in the text that follows.

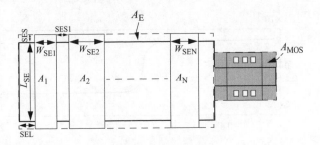

Figure 2.20   Definition of the parameters in eq. (2.18)

The total area of the transistor (A_T) has two main components: the area of the MOS device (A_{MOS}) and the area occupied by the new effective inputs, that corresponds to the extra surface of FG outside the active area (A_E):

$$A_T = A_{MOS} + A_E \tag{2.17}$$

The component A_E will vary with the number of inputs N in the FGMOS:

$$A_E = \sum_{i=1}^{N} A_i + \text{SES1} \cdot (N-1) \cdot L_{SE} + 2 \cdot \text{SES} \sum_{i=1}^{N} W_{SEi}$$
$$+ 2 \cdot \text{SES1} \cdot \text{SES} \cdot (N-1) + 4 \cdot \text{SEL} \cdot \text{SES} \tag{2.18}$$

where A_i, W_{SEi} and L_{SE} are the area, width and length of each effective input electrode for $i = [1, N]$, respectively, and SES1, SES and SEL are dimensions given by technological rules (Fig. 2.20). It can be proven, using eq. (2.18), that the extra area A_E will be minimised when the length and width of the input capacitors are chosen according to the following equations:

$$L_{SE} = \sqrt{\frac{2 \cdot \text{SES}}{\text{SES1}(N-1)} \sum_{i=1}^{N} A_i}, \quad W_{SEi} = \frac{A_i}{\sqrt{(2 \cdot \text{SES}/\text{SES1}(N-1)) \sum_{i=1}^{N} A_i}}$$

$$\tag{2.19}$$

The total extra area will then be

$$A_E = \sum_{i=1}^{N} A_i + 2 \sqrt{2 \cdot \text{SES} \cdot \text{SES1} \cdot (N-1) \sum_{i=1}^{N} A_i}$$
$$+ 2 \cdot \text{SES1} \cdot \text{SES} \cdot (N-1) + 4 \cdot \text{SEL} \cdot \text{SES} \tag{2.20}$$

Although in the minimum area solution the number of inputs is also minimum, sometimes, when matching is important it is more convenient to implement capacitances of different values by shorting capacitors with the same area. In this case, the minimum extra area required is

$$A_E = N \cdot A_1 + 2\sqrt{2 \cdot \text{SES} \cdot \text{SES1} \cdot (N-1) \cdot N \cdot A_1}$$
$$+ 2 \cdot \text{SES1} \cdot \text{SES} \cdot (N-1) + 4 \cdot \text{SEL} \cdot \text{SES} \tag{2.21}$$

where A_1 is the area of the unit capacitor. Matching is mostly an issue between capacitances that are used to process signals. Generally, the ratios between these capacitances are integer numbers, whereas the values of the capacitors which are used to bias or shift the signal levels do not keep an integer ratio. In such a case, the unit capacitance (capacitance that is to be replicated) should be the minimum value capacitance used to process the signal. For example, let us consider an FGMOS transistor with three inputs and assume that due to design requirements two of these inputs are connected to internal circuit nodes that vary with time and the weight of one of them is two times larger than the weight of the other. Let us also assume that the third input is connected to a constant voltage, and the ratio between its weight and the smaller one of the other two is 1.3. The recommended layout for this transistor would consist of four pieces of poly2 placed on top of the poly1 that forms the FG with relative widths (1,1,1,1.3). The weight 2 would be implemented by connecting two capacitances of weight 1 together.

The absolute value of the minimum capacitance (C_1) depends on the dimensions of the MOS transistor. There are two reasons for this.

1. The parasitic capacitances C_{GD}, C_{GS} and C_{GB} (that depend on the size of the transistor) reduce the value of the weights due to the effect they have on C_T (eq. (2.3)). This has direct consequences on the effective transconductance of the device (eq. (2.10)) and hence on the circuit speed. Therefore, during the design process, once it has been decided about the number of device inputs as well as its minimum acceptable transconductance and hence the minimum w_1 weight, that depends on the minimum area A_1, the implemented weight will deviate from the ideal value by a percentage:

$$\varepsilon \approx \frac{100 \cdot (C_{GS} + C_{GD} + C_{GB})}{\sum_{i=1}^{N} C_i} = \frac{100 \cdot (C_{GS} + C_{GD} + C_{GB})}{A_1 C_{SiO_2} \sum_{i=1}^{N} (A_i/A_1)} \tag{2.22}$$

where C_{SiO_2} is the capacitance per unit area of the poly2 on top of the poly1 with SiO_2 in between, that forms the effective inputs and the FG, respectively ($\varepsilon_{SiO_2}/t_{SiO_2}$).

2. The second reason is related to the way in which the current in the FGMOS transistor is affected by the ratio between C_{GD} and C_T. When a transistor is operating in saturation region (either in weak or strong inversion), a high output resistance is desired. However, in general, in FGMOS devices, the value of the output conductance is dominated by the C_{GD}/C_T term in eq. (2.11). If this term is not small enough it could cause unwanted effects such as distortion or gain loss. These effects could be minimised by keeping V_{DS} constant which can be achieved by using some circuit techniques such as cascoding. However, the term $C_{GD}V_{DS}/C_T$ could still be a problem because of the shift of the operating point at the FG with respect to the weighted sum of the voltages at the effective inputs (see eq. (2.2)).

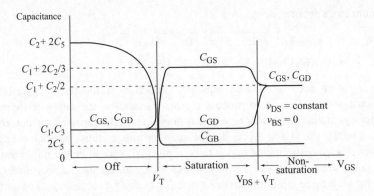

Figure 2.21   Capacitances C_{GS}, C_{DS} and C_{GB} as a function of V_{GS} (V_{DS} = constant and V_{BS} = 0) [119]

C_{GD} and C_{GS} have two contributors: the overlap capacitance due to the overlap between the gate and drain/source, respectively; and the intrinsic capacitance which depends on the operating mode of the device. C_{GB} has three contributors: the overlap capacitance, which occurs between the gate and bulk at the edges of the channel; the intrinsic charge-storage capacitance; and the parasitic capacitance between the portion of poly1 layer outside the channel that constitutes the major part of the FG and the substrate (proportional to A_E). The first two components are common to MOS devices. The third component is typical of FGMOS transistors and, in general, is much larger than the others. Figure 2.21 shows how C_{GD} and C_{GS} vary as the transistor enters different operating regions. A simplified mathematical model for C_{GD}, C_{GS} and C_{GB} is shown in eq. (2.23) to eq. (2.31) [119]. This model does not predict the smooth transitions as shown in Fig. 2.21 but is good enough for our derivations.

Weak inversion saturation

$$C_{GB} = C_{ox} \cdot W \cdot L + \text{CGBO} \cdot L + \frac{\varepsilon_{SiO_2}}{t_{SiO_2}^F} \cdot A_E \tag{2.23}$$

$$C_{GS} = \text{CGSO} \cdot W \tag{2.24}$$

$$C_{GD} = \text{CGDO} \cdot W \tag{2.25}$$

Strong inversion saturation

$$C_{GB} = \text{CGBO} \cdot L + \frac{\varepsilon_{SiO_2}}{t_{SiO_2}^F} \cdot A_E \tag{2.26}$$

$$C_{GS} = \text{CGSO} \cdot W + (2/3)(C_{ox} \cdot W \cdot L) \tag{2.27}$$

$$C_{GD} = \text{CGDO} \cdot W \tag{2.28}$$

Strong inversion ohmic

$$C_{GB} = CGBO \cdot L + \frac{\varepsilon_{SiO_2}}{t^F_{SiO_2}} \cdot A_E \tag{2.29}$$

$$C_{GS} = CGSO \cdot W + 0.5 \cdot C_{ox} \cdot W \cdot L \tag{2.30}$$

$$C_{GD} = CGDO \cdot W + 0.5 \cdot C_{ox} \cdot W \cdot L \tag{2.31}$$

Parameters CGBO, CGSO and CGDO depend on the technology and account for the overlap capacitances per unit of effective width (or length). C_{ox} is the value of the capacitance per unit of area in the active region ($\varepsilon_{SiO_2}/t_{ox}$) and $t^F_{SiO_2}$ is the thickness of the field oxide. For example, for a transistor operating in the strong inversion saturation region the deviation of the value of the minimum w_1 weight with respect to the ideal value will depend on the transistor dimensions and technology parameters in the following way:

$$\varepsilon \approx \frac{100 \cdot (CGBO \cdot L + CGSO \cdot W + CGDO \cdot W + (2/3)(C_{ox} \cdot W \cdot L) + (\varepsilon_{SiO_2}/t^F_{SiO_2})A_E)}{C_{SiO_2} A_E} \tag{2.32}$$

Hence, if for a device with N inputs ε has to be smaller than a certain value, e, due to design constraints such as a minimum allowed effective transconductance, the smallest extra area required, A_E^{min} would be

$$A_E^{min} = \frac{(CGBO \cdot L + CGSO \cdot W + CGDO \cdot W + (2/3)(C_{ox} \cdot W \cdot L))}{(0.01 \cdot e \cdot C_{SiO_2} - \varepsilon_{SiO_2}/t^F_{SiO_2})} \tag{2.33}$$

As an example, let us consider an FGMOS transistor in a typical technology with three identical effective inputs, effective width and length W = 10 μm, L = 10 μm, respectively, and CGBO = 3.45×10^{-10} F/m, CGDO = 1.38×10^{-10} F/m, CGSO = 1.38×10^{-10} F/m, $t_{ox} = 10^{-8}$ m and $t^F_{SiO_2} = 310 \times 10^{-10}$ m. If the maximum desired variation of the minimum weight with respect to its ideal value is 15 per cent, the minimum extra area would be 2130 μm^2.

If ε is not an issue, but the value of the output resistance, which is normally the case, then, the term C_{GD}/C_i has to be smaller than a certain value, k'. In this situation, a design equation that can be used to estimate the area is

$$\frac{C_{GD}}{C_i} \approx \frac{N \cdot CGDO \cdot W}{C_{SiO_2} \cdot A_E} < k' \Rightarrow A_E^{min} = \frac{N \cdot CGDO \cdot W}{C_{SiO_2} \cdot k'} \tag{2.34}$$

In order to illustrate this, let us assume the same device as in the previous example. If the output resistance of the MOS transistor that is used to build the FGMOS is sufficiently high, as to neglect the first term in eq. (2.11), and the effective transconductance is required to be at least 100 times larger than the output conductance, then the minimum extra area required would be 372 μm^2.

2.5   Summary and conclusions

The FGMOS transistor can, in general, be treated as a normal MOS transistor. It can be fabricated in any MOS technology, although for a good performance, a double polysilicon layer is recommended. The physical models used to describe the MOS transistor can be adapted for the FGMOS just by applying a change of variables into the equations. Similarly, the same simulation models can be used, as long as the simulations are set up properly.

The main disadvantages of the FGMOS are the reduction of the output resistance and the effective transconductance, the increased area and the uncertain amount of charge that might remain trapped at the FG during fabrication. However, with a careful design and layout, the extra area can be minimised and the effect of the charge made negligible. The reduction of the output resistance can be compensated for with design techniques that will be discussed in the following chapters.

Notation

ε	Deviation between the implemented and the ideal weight in FGMOS device (eq. (2.22))
ε_{SiO_2}	Dielectric constant of the SiO_2
A_1	Area of the unit capacitor (eq. (2.21))
A_E	Area occupied by new effective inputs in FGMOS (extra surface of FG outside the active area) (eq. (2.18), Fig. 2.20)
A_E^{min}	The smallest extra area required for an FGMOS device with N inputs and ε (eq. (2.22)) smaller than a certain value e (eq. (2.33))
A_i for $i = [1, N]$	Area of the i-th effective input electrode (Fig. 2.20)
a_i; a_o; a_o'	Gains of VCVSs in Fig. 2.5
A_{MOS}	Area of the MOS device in an FGMOS transistor (eq. (2.17), Fig. 2.20)
$A_T : A_{MOS} + A_E$	Total area of an FGMOS transistor (eq. (2.17), Fig. 2.20)
C_1	Value of the minimum input capacitance (see Section 2.5)
C_{GB}	Gate-to-bulk parasitic capacitance
CGBO	Overlap bulk capacitance per unit of length
C_{GD}	Gate-to-drain parasitic capacitance
CGDO	Overlap drain capacitance per unit of width
C_{GS}	Gate-to-source parasitic capacitance
CGSO	Overlap source capacitance per unit of width
C_i for $i = [1, N]$	i-th FGMOS input capacitance
C_{ox}	Capacitance per unit of area in the active region ($\varepsilon_{SiO_2}/t_{ox}$)
C_{SiO_2}	Capacitance per unit area of the poly2 on top of the poly1 with SiO_2 in between that form the effective inputs and the FG, respectively ($\varepsilon_{SiO_2}/t_{SiO_2}$) (eq. (2.22))
C_T	Total capacitance seen by the FG (eq. (2.3))

C_T'	Total capacitance seen by gate of the quasi-FGMOS in Fig. 2.17 (different from C_T in the normal FGMOS due to R_{leak})
FG	Floating Gate
G	Gain of the common source amplifier in Fig. 2.4(a)
G_{FG}	Gain of the common source amplifier in Fig. 2.4(b)
g_{ds}	MOS transistor output conductance
$g_{ds}^{(i)}$ for $i = [1, 3]$	Mi output conductance (Fig. 2.4)
g_{dsF}	FGMOS transistor output conductance (eq. (2.11))
g_m	MOS transistor gate transconductance
$g_m^{(i)}$ for $i = [1, 3]$	Mi transconductance (Fig. 2.4)
g_{mb}	MOS transistor bulk transconductance
g_{mbF}	FGMOS transistor bulk transconductance (eq. (2.12))
g_{mi} for $i = [1, N]$	Effective i-th input transconductance of an FGMOS with N inputs (eq. (2.10))
k	Number of metals in the technology
k'	Minimum value of C_{GD}/C_i required by design constraints (eq. (2.34))
L	Effective length of transistor
L_{SE}	Length of the i-th effective input electrode (the same length has been assumed for all inputs) (Fig. 2.20, eq. (2.19))
Q_{FG}	Charge trapped in the FG during fabrication
R_G	Resistance connected to the FG in Ramirez-Angulo's model (Fig. 2.5)
R_{leak}	Quasi-FGMOS resistance implemented using active devices (Fig. 2.17)
R_{out}	Output resistance in common source amplifier in Fig. 2.4(a)
R_{outFG}	Output resistance of the common source amplifier in Fig. 2.4(b)
SEL	Dimension given by technological rules – related to extra surface of FG (Fig. 2.20)
SES	Dimension given by technological rules – related to extra surface of FG (Fig. 2.20)
SES1	Dimension given by technological rules – related to extra surface of FG (Fig. 2.20)
t_{SiO_2}	Thickness of the SiO_2 between the FG and the effective inputs V_i
$t_{SiO_2}^F$	Thickness of the field oxide below the FG outside the active area
V_{FG}	Voltage at the FG
$v(f_{gi})_{op}$	Operating point at the FG for simulation purposes (Fig. 2.7)
v_i for $i = [1, N]$	FGMOS effective inputs
$v_{in}^2/\Delta f|_{\text{effective input}}$	Noise power spectral density at the effective input (eq. (2.15))

$\overline{v_{\text{in}}^2}/\Delta f\vert_{\text{FG}}$	Noise power spectral density at the FG (eq. (2.15))
V_{T}	Threshold voltage (Table 2.1)
W	Transistor effective width
$W_{\text{SE}i}$ for $i = [1, N]$	Width of the i-th effective input electrode (Fig. 2.20, eq. (2.19))
w_1	The minimum weight in an FGMOS – depends on the minimum area A_1 of its corresponding input electrode
w_i for $i = [1, N] : C_i/C_{\text{T}}$	FGMOS equivalent i-th input weight

Chapter 3

FGMOS – Circuit applications and design techniques

3.1 Introduction

This chapter explains the advantages of using an FGMOS device over a standard MOS transistor in circuits which require low power and low voltage operation. The first section introduces three circuit equivalents for the FGMOS device which are then used in subsequent chapters to design more complex circuits with a higher functionality. Several simple circuits are then presented in order to give the reader a flavour of various new design paths that the FGMOS transistor opens, in addition to illustrating the device functionality. The design tradeoffs are also explained together with examples. Novel, state-of-the-art building blocks and architectures can then be developed from the presented structures. They are left to the reader to optimise and refine. More advanced circuits are introduced and analysed in subsequent chapters.

3.2 Initial design ideas: three circuit equivalents for the FGMOS

This section introduces three circuit equivalents for the FGMOS transistor: an adder; a variable V_T FET and a current multiplier, which present the device as a tunable, controllable and flexible element. Seeing the FGMOS as one of these equivalent blocks can help the designer to anticipate when the transistor might be useful in a specific design.

3.2.1 FGMOS: a MOS transistor plus an adder

As eq. (2.2) shows, the FGMOS could be seen as a normal MOS transistor with a weighted sum of voltages applied to its gate (Fig. 3.1). The weights are given by the

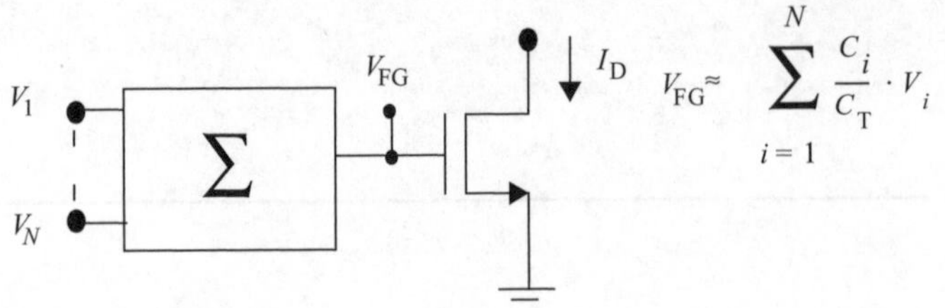

Figure 3.1   Equivalent circuit for the FGMOS

ratios between each input capacitance and the total capacitance seen by the FG. Since the addition can be performed without having to use any extra circuitry other than a number of capacitors, the use of this device can simplify certain circuit topologies and have very positive implications on the circuit in terms of area, noise, DR and power consumption. Besides, simpler designs can be of crucial importance in circuits that have to operate by pushing the limits of the technology. For example, if a circuit requiring addition has to operate at a maximum voltage supply well below the technological limit, not having to design an analog adder block can make the design process easier.

3.2.2   FGMOS: a MOS transistor with a controllable threshold voltage

The FGMOS device can also be seen as a MOS transistor with a controllable threshold voltage. In order to understand this, let us rewrite the equation for the current in an n-channel FGMOS transistor, operating in the strong inversion saturation region (eq. (2.9))[6] assuming as a first approximation that the intrinsic parasitic capacitances are much smaller than the input capacitances:

$$
I_D = \beta_{FG} \left[V_{iS} - \left(\frac{C_T}{C_i} V_T - \sum_{\substack{j=1 \\ j \neq i}}^{N} \frac{C_j}{C_i} V_{jS} - \frac{C_{GD}}{C_i} V_{DS} - \frac{C_{GB}}{C_T} V_{BS} - \frac{Q_{FG}}{C_i} \right) \right]^2
$$

$$
\approx \beta_{FG} \left[V_{iS} - \left(\frac{C_T}{C_i} V_T - \frac{C_{GB}}{C_i} V_{BS} - \sum_{\substack{j=1 \\ j \neq i}}^{N} \frac{C_j}{C_i} V_{jS} - \frac{Q_{FG}}{C_i} \right) \right]^2
$$

$$
= \beta_{FG} \left(V_{iS} - \frac{C_T}{C_i} V_{TFG} \right)^2
$$

$$
= \beta_{FG} (V_{iS} - V_T')^2 \tag{3.1}
$$

---

[6] A similar derivation can be carried out for other operating regions.

where

$$\beta_{FG} = \frac{\mu_o C_{ox}}{2} \frac{W}{L} \left(\frac{C_i}{C_T}\right)^2 = \frac{\mu_o C_{ox}}{2} \frac{W}{L} w_i^2 \tag{3.2}$$

$$V_T' = \frac{C_T}{C_i} V_T - \frac{C_{GB}}{C_i} V_{BS} - \sum_{\substack{j=1 \\ j \neq i}}^{N} \frac{C_j}{C_i} V_{jS} - \frac{Q_{FG}}{C_i} = \frac{C_T}{C_i} V_{TFG} \tag{3.3}$$

The equation shows how the FGMOS behaves as an MOS transistor with a β_{FG} parameter that depends on the weight w_i at the effective input (eq. (3.2)), and a threshold voltage parameter, V_T' (eq. (3.3)) that can be controlled electrically by changing the voltages at the additional inputs (V_{jS}, for $j = [1, N]$ ($j \neq i$)). This is one of the biggest advantages of the FGMOS transistor as it enables programming of the signal levels individually in each device according to the needs of the specific circuit. Besides, eq. (3.3) shows that it is possible to reduce the effective threshold voltage, make it zero or even invert its sign. This makes FGMOS devices ideal for implementation of very low voltage circuits, or circuits that require a high degree of tunability/programmability. Hence, they can, for example, compensate for big variations in performance due to mismatch.

3.2.3   FGMOS in weak inversion: a current multiplier

This section will explain how an FGMOS transistor biased in the weak inversion saturation region can behave as a nonlinear circuit block and perform multiplication of rational powers of currents. In order to understand this, let us start by considering a MOS transistor operating in the weak inversion saturation region with a current I_i flowing through it. From eq. (2.4), if $V_{DS} > 4U_T$ and $V_{BS} = 0$

$$I_i = I_s e^{V_{GS}/nU_T} \tag{3.4}$$

And if the gate voltage is V_i

$$e^{V_{is}/nU_T} = \frac{I_i}{I_s} \tag{3.5}$$

A similar expression can be obtained for N equally sized MOS transistors with currents I_j (for $j = [1, N]$) and gate voltages V_j (for $j = [1, N]$).

Let us now consider an FGMOS transistor with the same channel size, same voltages, V_j for $j = [1, N]$, at its effective inputs and also the same voltage V_S at the source, as in the case of the N equally sized MOS transistors. Using eq. (2.5) the current through this FGMOS transistor will be

$$I_D = I_S \prod_{j=1}^{N} \left[\exp\left(\frac{V_{jS}}{nU_T}\right)\right]^{C_j/C_T} \exp\left(\frac{C_{GD}}{C_T}\frac{V_{DS}}{nU_T}\right) \times \exp\left(\frac{Q_{FG}}{C_T nU_T}\right) \tag{3.6}$$

The voltages V_j can be regarded as voltages generated at the gates of the N MOS transistors by their respective drain currents I_j (for $j = [1, N]$). Equation (3.6) can be

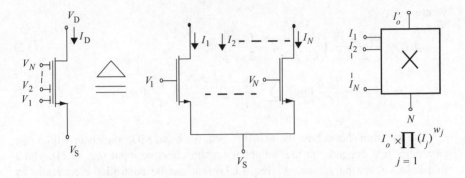

Figure 3.2   Equivalent circuit for the FGMOS in weak inversion: a current multiplier

rewritten as a function of these currents:

$$I_D \approx I_s \times \prod_{j=1}^{N} \left(\frac{I_j}{I_s}\right)^{w_j} \times \exp\left(\frac{Q_{FG}}{C_T n U_T}\right) \times \exp\left(\frac{C_{GD} V_{DS}}{C_T n U_T}\right) \tag{3.7}$$

where $w_j = C_j/C_T$. Assuming that the gate-to-drain parasitic capacitance is much smaller than the input capacitances and hence $(C_{GD}V_{DS}/C_T n U_T) \ll 1$, eq. (3.7) can be further simplified resulting in the expected nonlinear relationship between currents:

$$I_D \approx I_0' \times \prod_{j=1}^{N} (I_j)^{w_j} \tag{3.8}$$

where $I_0' = I_s^{(1-\sum_{j=1}^{N} w_j)} \times \exp(Q_{FG}/C_T n U_T)$.

Therefore, the FGMOS transistor working in the subthreshold saturation region could be seen as a multiplier of currents with rational or integer exponents. The currents being multiplied are those flowing through N different MOS transistors that have (a) the same source voltage as the FGMOS and (b) their gates voltages are the same as the voltages at the N FGMOS effective inputs. The equivalent circuit as well as the schematic symbol for the FGMOS operating as a current multiplier is shown in Fig. 3.2.

3.3   Circuits applications and design techniques

The aim of this section is to illustrate the performance of FGMOS devices using very simple circuit blocks. The design procedure is explained so that the reader can apply it to the design of other circuits with FGMOS.

3.3.1   A cascode current mirror

One of the simplest circuits in which replacing an MOS transistor by an FGMOS can be beneficial is current mirrors. This would be used, for example, when the power

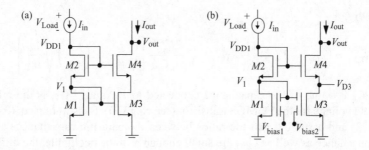

Figure 3.3   *(a) Conventional cascode mirror (b) Cascode current mirror with FGMOS*

supply voltage is one of the main design constraints. In order to show this, let us consider the current mirrors in Fig. 3.3. Figure 3.3(b) is the FGMOS realisation of the MOS current mirror in Fig. 3.3(a).

The bottom MOS transistors have been replaced by FGMOS devices with two effective inputs. One of the inputs acts as the MOS gate in the conventional MOS current mirror, whereas the other is connected to a constant bias voltage, V_{bias1} for device M1 and V_{bias2} for M3. Let us assume that $V_{bias1} = V_{bias2} = V_{DD}$, where V_{DD} is the maximum supply voltage. The voltage V_{DD1} can be written as a function of M2 gate-to-source voltage, V_{GS2}, and M1 drain to source voltage, V_1 as

$$V_{DD1} = V_{GS2} + V_1 \tag{3.9}$$

For a proper operation of the current mirrors, the transistors should operate in the strong inversion saturation region. Using eq. (2.8), eq. (3.9) can be rewritten as a function of the current

$$V_{DD1} = V_T + \sqrt{\frac{2I_{in}L_2}{\mu_o C_{ox} W_2}} + V_1 \tag{3.10}$$

where W_i and L_i are the effective width and length of transistor Mi, respectively. As a first approximation all threshold voltages are assumed to have the same value, V_T. Hence, the minimum voltage supply V_{DD}^{min} the current mirror would need to drive a maximum input current I_{in}^{max} would be

$$V_{DD}^{min} = V_{Load} + V_T + \sqrt{\frac{2I_{in}^{max}L_2}{\mu_o C_{ox} W_2}} + V_1 \tag{3.11}$$

where V_{Load} is the voltage across in the biasing/input part of the circuit (I_{in}). Equation (3.11) can now be rewritten for both MOS and FGMOS current mirrors:

MOS

$$V_{DD}^{min} = V_{Load} + 2V_T + \sqrt{\frac{2I_{in}^{max}}{\mu_o C_{ox}}} \left(\sqrt{\frac{L_2}{W_2}} + \sqrt{\frac{L_1}{W_1}} \right) \tag{3.12}$$

FGMOS

$$V_{DD}^{min} = V_{Load} + V_T + \frac{C_T}{C_1} V_{TFG} + \sqrt{\frac{2I_{in}^{max}}{\mu_o C_{ox}}} \left[\sqrt{\frac{L_2}{W_2}} + \left(\frac{C_T}{C_1}\right) \sqrt{\frac{L_1}{W_1}} \right] \quad (3.13)$$

where C_1 is value of the capacitance connected to V_1, and, V_{TFG} is the effective threshold voltage for the FGMOS transistor (see eq. (3.3)). Substituting eq. (3.3) into eq. (3.13) and assuming that the ratios between the parasitic capacitances and the total capacitance as well as Q_{FG} are small enough as to be negligible, the difference between the minimum voltage supply needed by these two circuit cells is

$$\Delta V_{DD} \approx \frac{1}{(1 + C_1/C_{VDD})} \left[V_{Load} + V_T + \sqrt{\frac{2I_{in}^{max} L_2}{\mu_o C_{ox} W_2}} \right]$$

$$= \frac{1}{(1 + C_1/C_{VDD})} \left[V_{Load} + V_T + \sqrt{\frac{I_{in}^{max}}{\beta_2}} \right] \quad (3.14)$$

where C_{VDD} is the capacitance connected to V_{DD} and β_i is the β parameter for transistor Mi. Equation (3.14) represents how much smaller the minimum voltage supply can be for an FGMOS current mirror than for its equivalent MOS current mirror. Hence, the constraint for the minimum supply voltage can be less restrictive in an FGMOS design if the capacitances ratios are adequately chosen for a given current range and aspect ratios.

Hence, for example, if the current mirrors in Fig. 3.3 are designed in a 0.35 μm technology, the minimum value of the power supply voltage that the MOS current mirror needs for the correct operation is 1.15 V, whereas the FGMOS current mirror works at 0.85 V. The values of the input capacitances used for the FGMOS current mirror are given in Table 3.1.

Figure 3.4 shows the input–output characteristics of both the MOS and the FGMOS current mirrors operating at a power supply voltage of 0.85 V. The figure shows how the output current of the FGMOS current mirror perfectly follows the

Table 3.1   *Design parameters for the FGMOS current mirror in Fig. 3.3*

Device	Parameter	Value (fF)
M1	C_1	225
M1	C_{VDD}	175
M2	C_1	225
M2	C_{VDD}	175

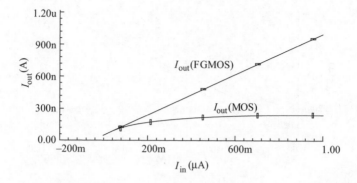

Figure 3.4   Input–output characteristic of the cascode current mirrors in Fig. 3.3 with $V_{DD} = 0.85$ V. The bottom line is the output current of the MOS cascode current mirror. The top line is the output current of the FGMOS current mirror

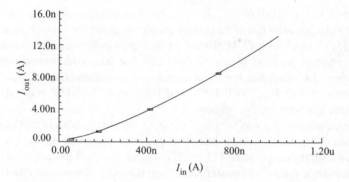

Figure 3.5   Output current versus input current for the FGMOS cascode current mirror in Fig. 3.3(b) for a mismatch of 0.194 V in threshold voltages of the FGMOS transistors, when $V_{bias1} = V_{bias2} = 0.85$ V

input current along the whole range of input values, whereas the output of the MOS current mirror saturates at around 140 nA.

In addition to reducing the minimum voltage requirement, FGMOS transistors can also be beneficial for a different reason. In general, they can be very useful in circuits fabricated in technologies that cannot guarantee a good level of matching between adjacent devices (an example could be polysilicon thin film transistors (TFT) technologies in which the variations between the threshold voltage of two identical devices can be worse than 1 V). In order to illustrate this let us assume that the difference between the threshold voltages of the bottom transistors in Fig. 3.3 is 0.194 V. Figure 3.5 shows the input–output characteristic for the FGMOS current mirror in this situation. In the MOS circuit it would be impossible to compensate for this mismatch by using external tuning signals. In the FGMOS, however, the mismatch can be corrected by applying different bias signals, V_{bias1} and V_{bias2}, to the

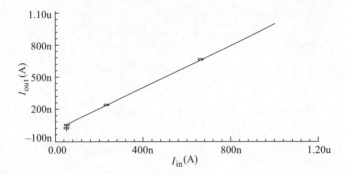

Figure 3.6   Output current versus input current for the FGMOS cascode current mirror in Fig. 3.3(b) for a mismatch of 0.194 V between the threshold voltages of the FGMOS transistors, when $V_{bias1} = 0.425$ V and $V_{bias2} = 0.85$ V

C_{VDD} inputs in the FGMOS pair. The mismatch is corrected because the effective threshold voltages of FGMOS transistors are functions of the voltage values at the extra device inputs (see eq. (3.3)). Hence, by creating a difference between the values of the voltages at the extra inputs the final effective threshold voltages will be the same. Figure 3.6 illustrates the input current-output current characteristic for the FGMOS circuit with $V_{bias1} = 0.425$ V and $V_{bias2} = 0.85$ V. The graph shows that the variation has been totally compensated.

   This example shows a way to 'adapt' the input–output characteristic of an FGMOS circuit after fabrication, without having to provoke tunnelling and/or hot electron injection within the circuit itself [73–81,92]. The challenge is to determine the compensation voltage for each transistor and design the extra circuitry required by every specific topology.

3.3.2   Two-stage OPAMP

This section describes how to add flexibility and reduce the overall power consumption in a CMOS operational amplifier by replacing some of the MOS transistors with FGMOS devices. The disadvantages of doing this are also discussed, taking as an example the circuit in Fig. 3.7.

   The two parameters that determine the power consumed by the circuit in Fig. 3.7 are the current flowing through the devices and the voltage supply needed for the circuit operation. A current reduction would affect negatively the performance of the circuit in terms of unity gain bandwidth (GBW) and slew rate (SR). If the voltage supply is decreased instead, the performance of the circuit (in terms of GBW and SR) would not be affected as long as the values of the currents are maintained together with the operating modes of the devices. Hence, the voltage supply would be the best variable to alter in order to reduce the power consumption. However, the voltage supply reduction (V_{DD} if $V_{SS} = 0$) usually results in an undesired behaviour of the devices as a consequence of their operation in non-saturated regions. Forcing the

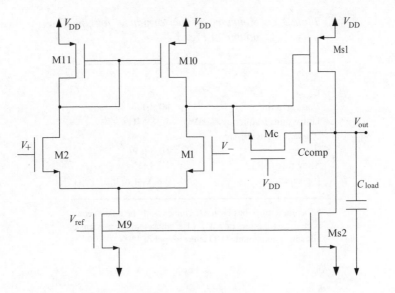

Figure 3.7   Two-stage compensated MOS opamp

transistors to work in strong inversion saturation region becomes impossible for a value of V_{DD} under a certain limit, and the only way to achieve this is to come up with new, innovative and complex circuit topologies. The use of FGMOS devices can enlarge the operative range for a lower value of V_{DD} without compromising the complexity of the circuits. In the opamp in Fig. 3.7, the minimum voltage supply is determined by two saturation voltages ($V_{DSAT(M9)} + V_{DSAT(M1/2)}$), the threshold voltage of a p-transistor ($|V_{TP}|$) and also, indirectly, by the common mode at the input, unless a level shifter is used. The option of adding a level shifter to change the common mode at the input is not always a good idea since it would increase the power consumption. Besides, the equivalent offset at the input would have an added component due to the offset of the level shifter. Hence, the common mode voltage indirectly sets a constraint to the minimum voltage supply if the voltage levels are wanted to be centred around $V_{DD}/2$, the input range maximised, and extra circuitry avoided.

In order to illustrate this, the two-stage opamp in Fig. 3.7 is designed in a 0.35 μm technology for the following specifications: a power consumption smaller than 12 μW, GBW larger than 0.8 MHz for a load of 5 pF, phase margin (PM) greater than 60°, gain over 60 dB and a maximised input range. The minimum voltage supply is in this case determined by the common mode input range that requires a minimum value of 1.5 V. The performance is summarised in Table 3.2.

In order to reduce the voltage supply and hence the power, the opamp is redesigned using FGMOS devices. A first possibility is to substitute the input transistors, M1 and M2 for two-input FGMOS transistors, as shown in Fig. 3.8. The values of the input capacitances for the new circuit are collected in Table 3.3. A summary of performance is shown in Table 3.4.

Table 3.2   Summary of performance for the
opamp in Fig. 3.7

V_{DD}^{min}	1.5 V
C_{comp}	3 pF
Voltage gain	80 dB
Unity gain frequency (GBW)	1.38 MHz
Phase margin (PM)	75°
Power	10.8 μW
Slew rate (SR)	1.2 MV/s
Input range[a]	0.8 V_{pp} = 0.53 V_{DD}

[a] The input range has been determined with the opamp connected as a buffer, and a 1 kHz sinusoidal input signal, allowing a maximum THD at the output of 1%.

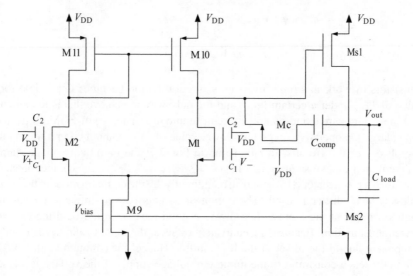

Figure 3.8   Two-stage FGMOS opamp: first version

Table 3.4 shows a 27 per cent reduction in the power consumption and almost 50 per cent improvement in the input range relative to V_{DD} for the FGMOS cell. However, the price to pay for this is a reduction of the gain and the GBW. The reduction of the GBW is due to the following: the GBW of the MOS opamp in Fig. 3.7 is given by[7]

$$GBW = \frac{g_{m1}}{C_{comp}}$$ (3.15)

---

[7] The subscripts refer to the name of the corresponding transistor.

Table 3.3 *Input capacitances for the FGMOS transistors in Fig. 3.8*

Device	Parameter	Value (fF)
M1	C_1	89.7
M1	C_2	175
M2	C_1	89.7
M2	C_2	175

Table 3.4 *Summary of the performance for the opamp in Fig. 3.8*

V_{DD}^{min}	1.1 V
C_{comp}	1.5 pF
Voltage gain	62 dB
Unity gain frequency (GBW)	0.84 MHz
Phase margin (PM)	$65°$
Power	7.9 µW
Slew rate (SR)	2.4 MV/s
Input range[a]	$0.8\ V_{pp} = 0.72\ V_{DD}$

[a] The input range has been determined with the opamp connected as a buffer, and a 1 kHz sinusoidal input signal, allowing a maximum THD at the output of 1%.

In the floating gate version of the opamp, using the expression for the effective input transconductance in eq. (2.10) together with eq. (3.15)

$$\text{GBW}_{FG} = \frac{C_1}{C_T} \frac{g_{m1}}{C_{comp}} \tag{3.16}$$

The compensation capacitance C_{comp} is two times smaller in the FGMOS version of the opamp. This on its own would double the GBW. However the ratio C_1/C_T is always smaller than one, in this case 0.3, which explains the overall reduction of the GBW by a factor of 1.6:

$$\frac{\text{GBW}_{FG}}{\text{GBW}} = \frac{C_1}{C_T} \cdot \frac{3}{1.5} \tag{3.17}$$

The effect on the gain is more serious. The reason for this is in equations (2.10) and (2.11). The total DC gain of the MOS amplifier (A_o) is

$$A_o = \frac{g_{m1}}{g_{ds1} + g_{ds10}} \times \frac{g_{ms1}}{g_{dss1} + g_{dss2}} \tag{3.18}$$

And for the FGMOS opamp

$$A_{oFG} = \frac{C_1}{C_T} \times \frac{g_{m1}}{g_{dsF1} + g_{ds10}} \times \frac{g_{ms1}}{g_{dss1} + g_{dss2}} \qquad (3.19)$$

The FGMOS reduces the gain because, on one hand the factor C_1/C_T is smaller than one and on the other hand, the output conductance of the input transistor is dominated by the term $C_{GD1}g_{m1}/C_T$ (see eq. (2.11)). Hence, in general, the gain is going to be limited to

$$A_{oFG} \approx \frac{C_1}{C_{GD1}} \times \frac{g_{ms1}}{g_{dss1} + g_{dss2}} \qquad (3.20)$$

unless the total capacitance is significantly increased or, design strategies such as cascoding are used.

Let us analyse now why a smaller compensation capacitance is needed. The increase of the output conductance in the first stage will also shift the first pole in a non-compensated design towards a higher frequency value. This, on its own, would be detrimental for the PM since the first pole would be closer to the second one. As a consequence of this effect only, a higher value of the compensation capacitance would be required. However, the reduced gain will have a completely opposite effect on the circuit. In the case of this specific design, the reduction of the gain compensates for the increase of the first pole frequency in terms of the PM. This is why a smaller compensation capacitance is needed in spite of the expected higher frequency of the first pole without compensation. Analytically, this can be explained using the following equation for the PM [119]:

$$PM \approx \pm 180° - atan\left(\frac{GBW}{|w_{p1}|}\right) - atan\left(\frac{GBW}{|w_{p2}|}\right) \qquad (3.21)$$

where w_{p1} and w_{p2} are the angular frequencies of the first and second poles, respectively. The frequency of the second pole is the same for both the MOS and FGMOS designs whereas w_{p1} is larger in the FGMOS circuit, and thus increases the denominator of the second term in eq. (3.21). However this is compensated by the reduction of the GBW in the FGMOS topology (eq. (3.16)).

In summary, the advantages of using FGMOS transistors at the input of the opamp circuit are the reduction of the power consumption as well as power supply voltage and, the increase of the input range that now goes almost rail to rail since the topology is operating at 1.1 V. In this particular case, the addition of FGs does not increase the area since the compensation capacitance is smaller. The disadvantages are smaller gain and GBW.

A design of the two-stage opamp in which a further reduction of the power supply voltage can be achieved is shown in Fig. 3.9. The idea is to shift the effective threshold voltages (eq. (3.3)) of all those transistors whose operating modes would be affected

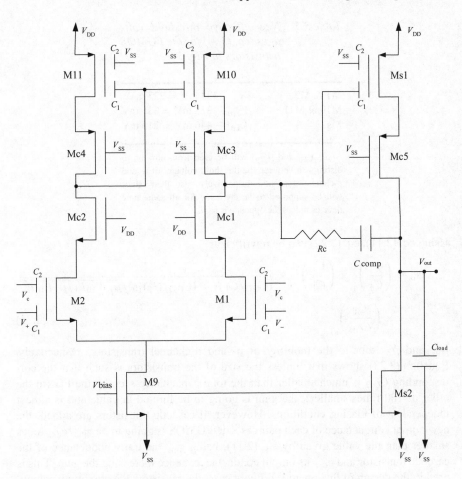

Figure 3.9   Two stage opamp with FGMOS: second version

by the reduction of the supply voltage. These are M1, M2, M10, M11 and Ms1 (Fig. 3.8). The new effective threshold voltages are shown in Table 3.5.[8]

The new opamp works with a power supply voltage of 0.9 V. In order to improve the performance, cascode transistors are added. The cascode transistors compensate for the degradation in output resistance and hence in gain caused by the $(C_{\text{GD}}g_\text{m})/C_\text{T}$ term added to the output conductances of the FGMOS transistors. Without cascode transistors, the gain of the amplifier would be[9]

$$A_{\text{oFG}} = \left(\frac{C_1}{C_\text{T}}\right)_1 \times \left(\frac{C_1}{C_\text{T}}\right)_{\text{s1}} \frac{g_\text{ml}}{g_{\text{dsF1}} + g_{\text{dsF10}}} \times \frac{g_{\text{ms1}}}{g_{\text{dsFs1}} + g_{\text{dss2}}} \tag{3.22}$$

---

[8] The input connected to either $V_\text{C} = V_{\text{DD}}$ or V_{SS} is used for this purpose.

[9] The subscripts in the capacitances ratios refer to the names of the transistors they correspond to.

Table 3.5   New effective threshold volt-
ages (eq. (3.3)) for the FGMOS
transistors in Fig. 3.9

M1 & M2	$V_{Tn} - 334\,\mathrm{mV} = 225\,\mathrm{mV}$
M10 & M11	$\lvert V_{Tp}\rvert - 318\,\mathrm{mV} = 335\,\mathrm{mV}$
Ms1	$\lvert V_{Tp}\rvert - 446\,\mathrm{mV} = 213\,\mathrm{mV}$

Note: V_{Tn} and $\lvert V_{Tp}\rvert$ will be used from now on to distinguish between the threshold voltages of n- and p-channel transistors, respectively. Also, their values will be supposed to be the same for all same type devices unless the opposite is said.

Using eq. (2.11), eq. (3.22) can be rewritten as

$$A_o \approx \left(\frac{C_1}{C_T}\right)_1 \times \left(\frac{C_1}{C_T}\right)_{s1} \times \frac{1}{(C_{GD}/C_T)_1 + (C_{GD}/C_T)_{10}\sqrt{\mu_p W_{10}L_1/\mu_n L_{10}W_1}}$$

$$\times \left(\frac{C_T}{C_{GD}}\right)_{s1} \tag{3.23}$$

μ_p, and μ_n refer to the mobility of p- and n-channel transistors, respectively. Equation (3.23) shows that, unless the size of the transistors is such that the corresponding C_{GD} is much smaller than the total capacitance seen by the FG (in the order of 100 times smaller), the gain is going to be limited in value and is almost independent of biasing conditions. However, if cascode transistors are added, the new output conductance of each pair cascode/FGMOS is going to be g_{mc}/g_{dsc} times smaller than the value given by eq. (2.11), being g_{mc} the transconductance of the cascode transistor and g_{dsc} its output conductance, hence increasing the gain. This is used in the design of this opamp. Without cascode transistors the maximum achievable DC gain is in the order of 40 dB, whereas the addition of cascode transistors increases it to 78 dB. A summary of the performance for the amplifier in Fig. 3.9 is shown in Table 3.6. The values of the input capacitances are collected in Table 3.7.

As a summary of the above, FGMOS transistors add a number of advantages and disadvantages to the design. The decision to use or not to use them will ultimately depend on the design tradeoffs.

The negative points are

1. A larger compensation capacitance is needed due to the addition of the cascode transistors in the first stage. Since the gain of the latter increases the compensation capacitance has to be larger in order to achieve an acceptable PM.
2. The SR is reduced because of the larger compensation capacitance.

Nevertheless, the transistor also makes it possible to

1. Reduce the power supply voltage,
2. Reduce the power consumption,

Table 3.6 Summary of performance for the opamp in Fig. 3.9

V_{DD}^{min}	0.9 V
C_{comp}	5 pF
Rc	400 k Ω
Voltage gain	78 dB
Unity gain frequency (GBW)	800 kHz
Phase margin (PM)	65°
Power	7.9 μW
Slew rate (SR)	0.9 MV/s
Input range[a]	$0.77V_{pp} = 0.85V_{DD}$

[a] The input range has been determined with the opamp connected as a buffer, and a 10 Hz sinusoidal input signal, allowing a maximum THD at the output of 1%.

Table 3.7 Design parameters for the FGMOS opamp in Fig. 3.9

Device	Parameter	Value (fF)
M1 & M2	C_1	89
M1 & M2	C_2	320
M10 & M11	C_1	89
M10 & M11	C_2	300
Ms1	C_1	79
Ms1	C_2	320

3. Increase the gain,
4. Increase the input range and
5. Adjust externally the threshold voltage of each individual device by changing the bias voltage at the extra FGMOS input, if required. This can be used for the compensation of process variations since the devices can be tuned externally after fabrication. The kind of circuitry required for this purpose would depend on the specific topology and the type of variations. As an example, an offset at the output of the opamp in Fig. 3.9 could be corrected by applying two different biasing voltages (V_C) to the input FGMOS transistors. Another possibility would be to apply different biasing voltages to the input capacities C_2 in transistors M10 and M11 (in this case the opamp would have to be designed with two disconnected FGs for these transistors). A final possibility would be to change the biasing voltage (connected to C_2) in transistor Ms1.

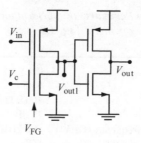

Figure 3.10   Two-input FGMOS inverter followed by a MOS inverter

3.3.3   FGMOS programmable inverter

This section describes how the flexibility/programmability of a CMOS inverter can be increased by replacing MOS with FGMOS transistors as shown in Fig. 3.10. The programmable FGMOS inverter can be used in a number of interesting applications that will be explained in subsequent chapters.

In a conventional CMOS inverter, the switching threshold U_{inverter} is given by

$$U_{\text{inverter}} \approx \frac{V_{\text{DD}} - |V_{\text{Tp}}| + \sqrt{\beta_n/\beta_p} \times V_{\text{Tn}}}{1 + \sqrt{\beta_n/\beta_p}} \qquad (3.24)$$

where again subscripts n and p refer to the n-channel and p-channel MOS transistors, respectively. If both, n-channel and p-channel MOS transistors are replaced with n-channel and p-channel 2-input FGMOS devices with common FG, and if one input of these new devices acts as the effective input and the other is connected to a bias voltage, the switching threshold would remain the same. This is, of course, assuming that the channel width and length of these transistors are the same as in the MOS implementation. However, in the new design, the value of the switching threshold does not have any practical meaning, since it represents the value of the voltage at which the FG and the output share the operating point ($V_{\text{FG}} = V_{\text{out1}}$), but the FG is not the input anymore. It is therefore, more appropriate to find the value of the voltage at the effective input for which this happens. This voltage can be obtained by combining eq. (2.2) and eq. (3.24), which yields

$$V_{\text{in(threshold)}} = \left(\frac{C_T}{C_{\text{in}}} - \frac{(C_{\text{GDn}} + C_{\text{GDp}})}{C_{\text{in}}} \right) \times \left(\frac{V_{\text{DD}} - |V_{\text{Tp}}| + \sqrt{\beta_n/\beta_p} \times V_{\text{Tn}}}{1 + \sqrt{\beta_n/\beta_p}} \right)$$
$$- \frac{C_c}{C_{\text{in}}} V_c - \frac{(C_{\text{GSp}} + C_{\text{GBp}})}{C_{\text{in}}} V_{\text{DD}} \qquad (3.25)$$

where C_c and C_{in} refer to the FGMOS capacitances connected to V_c and V_{in}, respectively. Equation (3.25) shows that it is possible to change the switching threshold of the FGMOS inverter just by varying the value of V_c. Experimental results for this circuit are shown in Fig. 3.11. V_{out1} represents the output of the FGMOS inverter obtained by sweeping V_{in} between 0 and 5 V and varying V_c in steps of one volt. V_{out} is the output of an MOS inverter that is connected in series with the FGMOS inverter

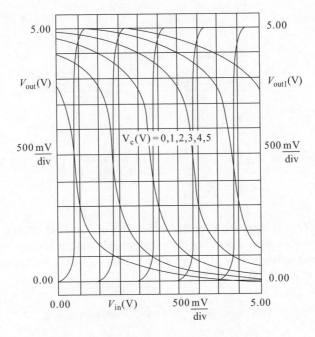

Figure 3.11   Input–output characteristic of inverter in Fig. 3.10

and whose transistors have the same aspect ratios as the corresponding devices in the FGMOS implementation. The purpose of the second inverter is twofold: on one hand it regenerates the digital levels; on the other hand, it allows to compare the performances of the MOS and FGMOS inverters as well as to estimate the matching. This is next explained by showing how this cell can be used to approximately determine the experimental value of Q_{FG}.

Ideally, as both inverters have the same aspect ratios their switching thresholds should also be the same. However, if a difference exists between them it would give us an idea of how much the threshold voltages of the FGMOS transistors vary with respect to the threshold voltages of their MOS counterparts. This would also be an indication of the possible amount of charge that remains trapped after fabrication.

Assuming that other sources of error are negligible[10], the variation of $U_{inverter}$ (eq. (3.24)) caused by threshold voltage variations would be

$$\Delta U_{inverter} \approx \frac{1}{1 + \sqrt{\beta_n/\beta_p}} \times \left(-\Delta |V_{Tp}| + \sqrt{\beta_n/\beta_p} \times \Delta V_{Tn} \right) \tag{3.26}$$

Equation (3.26) is valid for both, the MOS and the FGMOS inverter. However, again, in the case of the FGMOS inverter it should be rewritten referred to the effective

---

[10] This is justified by the fact that the main extra source of mismatch between FGMOS and MOS is Q_{FG}. If Q_{FG} is erased the mismatch would be of the same order of magnitude as in two identical MOS devices.

input, which gives

$$V_{\text{in(threshold)}} = \frac{C_T}{C_{\text{in}}} \cdot (U_{\text{inverter}} + \Delta U_{\text{inverter}}) - \frac{C_c}{C_{\text{in}}} V_c$$
$$- \frac{(C_{\text{GDn}} + C_{\text{GDp}})U_{\text{inverter}}}{C_{\text{in}}} - \frac{(C_{\text{GSp}} + C_{\text{GBp}})}{C_{\text{in}}} V_{\text{DD}} \qquad (3.27)$$

From eq. (3.27)

$$\Delta U_{\text{inverter}} = \frac{C_{\text{in}}}{C_T} \cdot \left(V_{\text{in(threshold)}} + \frac{C_c}{C_{\text{in}}} V_c \right) - U_{\text{inverter}} \cdot \left(1 - \frac{C_{\text{GDn}} + C_{\text{GDp}}}{C_T} \right)$$
$$+ \frac{(C_{\text{GSp}} + C_{\text{GBp}})}{C_{\text{in}}} V_{\text{DD}} \qquad (3.28)$$

Hence, if Q_{FG} is supposed to be the only source of threshold voltage variation in eq. (3.26) its value could be obtained from the output of this circuit, just by measuring the value of the voltage at the effective input when $V_{\text{out}} = V_{\text{in}} = U_{\text{inverter}}$ and using these values in eq. (3.28):

$$\frac{Q_{\text{FG}}}{C_T} \approx \frac{C_{\text{in}}}{C_T} \times \left(V_{\text{in(threshold)}} + \frac{C_c}{C_{\text{in}}} V_c \right) - U_{\text{inverter}} \times \left(1 - \frac{C_{\text{GDn}} + C_{\text{GDp}}}{C_T} \right)$$
$$+ \frac{(C_{\text{GSp}} + C_{\text{GBp}})}{C_{\text{in}}} V_{\text{DD}} \qquad (3.29)$$

As it was already mentioned in Chapter 2, in a design using metal contacts on the FG, the experimental value obtained for eq. (3.29) turns out to be of the same order of magnitude as the expected variations due to mismatch, so it is difficult to know, if this is due to (a) the simulated values of the couplings capacitances used to estimate Q_{FG}, which are different from the real ones; (b) mismatch between MOS transistors in the two inverters or (c) the still remaining charge. In any case, the variation is not critical and can be easily taken into account during the design process. It can be just treated as either normal mismatch or process variations.

3.3.4   *An offset compensated FGMOS comparator*

This section shows how the use of FGMOS transistors in a comparator circuit can (a) reduce the minimum required power supply voltage; (b) increase the input range and (c) compensate for offset variations.

Figure 3.12 shows a simplified schematic of the fully differential comparator. It consists of two n-channel FGMOS input transistors (M_1 and M_2) loaded by two cross coupled PMOS transistors (M_3 and M_4) that act as regenerative comparator elements. Each FGMOS device has three input gates. One of them represents the effective input and is connected to $V_{\text{in}}^{\pm}$ (the differential input is given as $V_{\text{ind}} = V_{\text{in}}^{+} - V_{\text{in}}^{-}$). A second input is connected to the comparator's clock V_{clk}. This input plays a double role: on one hand, when a comparison is being performed, the high value of the clock signal reduces the effective threshold voltage and so, the transistors turn ON even for low

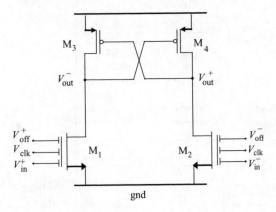

Figure 3.12   *FGMOS comparator schematic*

values of the input signals. On the other hand, transistors can be turned OFF when this input is low which will reset the comparator. Analytically this can be written as

$$\text{Input transistors ON if} : V_{in}^{+} > \frac{C_T}{C_{in}} \left(V_{Tn} - \frac{C_{clk}}{C_T} V_{clk} - \frac{C_c}{C_T} V_{off} \right) \qquad (3.30)$$

$$\text{Reset if} : V_{in}^{+} < \frac{C_T}{C_{in}} \left(V_{Tn} - \frac{C_c}{C_T} V_{off} \right) \qquad (3.31)$$

where C_{in} is the value of the capacitance connected to the effective input, C_{clk} is the value of the capacitance connected to the clock signal and C_c is the value of the capacitance connected to the offset compensation signal. The third input is therefore used for offset compensation. For the sake of simplicity, V_{off} has been assumed to take the same value for both differential branches ($V_{off}^{+} = V_{off}^{-} = V_{off}$), which is true in the absence of offset. However, in reality it will take a different value for each branch, which will consequently give rise to two different equations for eq. (3.30) as well as for eq. (3.31). This will be explained in more detail later on. Equations (3.30) and (3.31) show that the adequate choice of the capacitances ratios makes it possible to turn the input transistors ON for the whole range of input values when V_{clk} is high or, to turn them off and reset the comparator when V_{clk} is low.

3.3.4.1   The offset compensation

The main sources of offset in the comparator in Fig. 3.12 are explained in this section along with the strategies used for its compensation.

1. *Mismatch in the threshold voltage of the input transistors*: One of the main drawbacks of FGMOS devices is that any variation of the threshold voltage in the MOS transistor used to build the FGMOS is multiplied by the term C_T/C_{in} at the effective input. As the term C_T/C_{in} is always greater than one, the effects of the equivalent variation will always be greater than in a normal MOS device. This is critical in this comparator topology since the mismatch between FGMOS input transistors affects

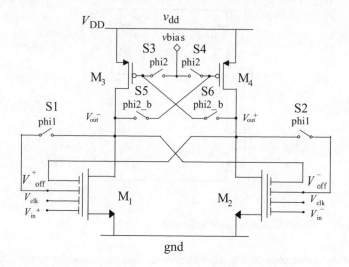

Figure 3.13   FGMOS comparator with offset compensation

the offset directly. Hence, if other sources of mismatch are ignored, the offset at the input caused by the variation between the threshold voltages of transistors M_1 and M_2, $\Delta V_T = (V_{Tn2} - V_{Tn1})$ (subscript i refers to device M_i), is

$$v_{\text{off}_{\Delta V_T}} = \frac{C_T}{C_{\text{in}}} \Delta V_T = \frac{C_T}{C_{\text{in}}} (V_{Tn2} - V_{Tn1}) \tag{3.32}$$

In order to compensate for this and other sources of offset the comparator in Fig. 3.12 is modified as shown in Fig. 3.13. The basic operation of the comparator is performed in two phases: *offset compensation phase and comparison phase* (Fig. 3.14). During the *offset compensation phase* switches S5 and S6 are opened while S1, S2, S3, S4 are closed. Switches S3 and S4 connect the gate of the load PMOS transistors to a constant voltage, V_{bias}, which makes them operate as current sources. As a first approach, ignoring the effects of mismatch between these two devices, the current flowing through them is

$$I_{\text{bias}} = \frac{\beta_p}{2} (V_{DD} - V_{\text{bias}} - |V_{Tp}|)^2 \tag{3.33}$$

Switches S1 and S2 establish a negative feedback by connecting one input of each of input FGMOS transistors with their respective drains (outputs of the comparator). The same value of the voltage, V_{cm}, is applied to V_{in}^+ and V_{in}^- during this phase. Assuming that the threshold voltages of both input transistors are different (V_{Tn1} for transistor M_1 and V_{Tn2} for transistor M_2), the voltage generated at $V_{\text{out}}^+ = V_{\text{off}}^-$ and

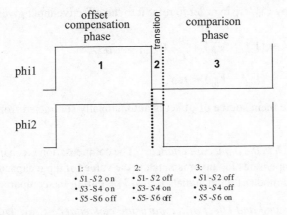

Figure 3.14 Timing diagram for the comparator in Fig. 3.13

$V_{out}^- = V_{off}^+$ by the negative feedback is

$$V_{off}^- = \frac{C_T}{C_c}\left(V_{Tn2} + \sqrt{\frac{2I_{bias}}{\beta_n}}\right) - \frac{C_{in}}{C_c}V_{cm} - \frac{C_{clk}}{C_c}V_{clk} \qquad (3.34)$$

$$V_{off}^+ = \frac{C_T}{C_c}\left(V_{Tn1} + \sqrt{\frac{2I_{bias}}{\beta_n}}\right) - \frac{C_{in}}{C_c}V_{cm} - \frac{C_{clk}}{C_c}V_{clk} \qquad (3.35)$$

During the comparison phase S1, S2, S3 and S4 are open (S1 and S2 must open slightly before S3 and S4 to avoid variation of V_{off}^- and V_{off}^+ during the transition), while S5 and S6 are closed. A positive feedback is thus created between the PMOS couple. A differential input voltage is applied to the input transistors. A small capacitance connected at nodes V_{off}^- and V_{off}^+ helps maintain the value of the voltages sampled during the compensation phase. Thus, based on (3.34) and (3.35), the voltage at the FGs of transistors M_1 and M_2 (V_{FG1} and V_{FG2}, respectively) will be

$$V_{FG1} = \frac{C_{in}}{C_T}(V_{in}^+ - V_{cm}) + V_{Tn1} + \sqrt{\frac{2I_{bias}}{\beta_n}} \qquad (3.36)$$

$$V_{FG2} = \frac{C_{in}}{C_T}(V_{in}^- - V_{cm}) + V_{Tn2} + \sqrt{\frac{2I_{bias}}{\beta_n}} \qquad (3.37)$$

Hence, the effective differential signal seen by the gates is

$$V_{FG1} - V_{FG2} = \frac{C_{in}}{C_T}(V_{in}^+ - V_{in}^-) + (V_{Tn1} - V_{Tn2}) \qquad (3.38)$$

And dividing by C_{in}/C_T in order to refer it to the effective input gives

$$V_{ind}^{(effective)} = (V_{in}^+ - V_{in}^-) + \frac{C_T}{C_{in}}(V_{Tn1} - V_{Tn2})$$

$$= (V_{in}^+ - V_{in}^-) - v_{off_{\Delta V_T}} \qquad (3.39)$$

In this way, the main source of offset is automatically subtracted from the effective input.

2. *Mismatch in the input capacitances*: The compensation mechanism described above also compensates for any mismatch in the values of input capacitances, as long as the common mode of input signals is also V_{cm} during the comparison phase.

3. *Feedthrough and effect of the parasitic capacitance C_{GD}*: During the comparison phase the voltages at nodes V_{off}^- and V_{off}^+ can vary with respect to the values they had at the end of the compensation phase. This could happen, for example, due to the feedthrough. In this case eq. (3.35) and eq. (3.36) are rewritten including these variations as

$$V_{FG1} = \frac{C_{in}}{C_T}(V_{in}^+ - V_{cm}) + V_{Tn} + \sqrt{\frac{2I_{bias}}{\beta_n}} - \frac{C_c}{C_T}\Delta V_{off}^+ \qquad (3.40)$$

$$V_{FG2} = \frac{C_{in}}{C_T}(V_{in}^- - V_{cm}) + V_{Tn} + \sqrt{\frac{I_{bias}}{\beta_n}} - \frac{C_c}{C_T}\Delta V_{off}^- \qquad (3.41)$$

where the terms ΔV_{off}^+ and ΔV_{off}^- account for them. The same threshold voltage is assumed for both input transistors in order to simplify derivations and also because the case with different threshold voltages has been previously discussed. Also, in the derivations above it has been assumed that the term $(C_{GD}/C_T)V_D$, in the expressions for the FGs voltages, is negligible, which is true only if $C_T \gg C_{GD}$. However, if the total capacitance is made too big, it will have a detrimental effect on both the area and speed of the circuit, and hence it will not represent a realistic case. Therefore, if the term corresponding to the parasitic coupling is included in previous derivations, the final expressions for (3.35) and (3.36), are

$$V_{FG1} = \frac{C_{in}}{C_T}(V_{in}^+ - V_{cm}) + V_{Tn} + \sqrt{\frac{2I_{bias}}{\beta_n}} - \frac{C_c}{C_T}\Delta V_{off}^+$$

$$+ \frac{C_{GD}}{C_T}[V_{out(comparison)}^- - V_{out(compensation)}^-] \qquad (3.42)$$

$$V_{FG2} = \frac{C_{in}}{C_T}(V_{in}^- - V_{cm}) + V_{Tn} + \sqrt{\frac{2I_{bias}}{\beta_n}} - \frac{C_c}{C_T}\Delta V_{off}^-$$

$$+ \frac{C_{GD}}{C_T}[V_{out(comparison)}^+ - V_{out(compensation)}^+] \qquad (3.43)$$

This on its own generates an equivalent input referred offset given by

$$v_{\text{off}(C_{\text{GD}}/\text{feedthrough})}$$

$$= \frac{C_{\text{GD}}}{C_{\text{in}}}[V^-_{\text{out(comparison)}} - V^+_{\text{out(comparison)}}]$$

$$- \frac{C_{\text{GD}}}{C_{\text{in}}}[V^-_{\text{out(compensation)}} - V^+_{\text{out(compensation)}}] - \frac{C_{\text{c}}}{C_{\text{in}}}(\Delta V^+_{\text{off}} - \Delta V^-_{\text{off}})$$

$$(3.44)$$

In general, the second and the third terms in (3.44) are negligible. The second term is proportional to a small variation, with a proportionality factor considerably smaller than one. In the third term, the proportionality factor $C_{\text{c}}/C_{\text{in}}$ does not need to be much smaller than one, since the variation due to the feedthrough is small. The first term, however, dominates as its absolute value is proportional to V_{DD}, that is, $(C_{\text{GD}}/C_{\text{in}})V_{\text{DD}}$, which is much larger, and thus, a compensation mechanism for this term is required. The compensation is performed by adding a small extra input, C_{f}, of value comparable with C_{GD} to M_1 and M_2 and applying a positive feedback by connecting it to the transistors opposite outputs (see Fig. 3.13). In this way, from (3.44) the new equivalent offset at the input is

$$v_{\text{off}(C_{\text{GD}}/\text{feedthrough})} \cong \frac{(C_{\text{GD}} - C_{\text{f}})}{C_{\text{in}}}[V^-_{\text{out(comparison)}} - V^+_{\text{out(comparison)}}] \qquad (3.45)$$

And this term can be made very close to zero, as proven by the design simulation results.

4. *Offset due to different threshold voltages in the PMOS*: Mismatch between the threshold voltages in the top PMOS transistors will result in different I_{bias} currents, I^+_{bias} and I^-_{bias}, flowing through the positive (M_1 and M_3) and negative (M_2 and M_4) comparator branches during the compensation phase. This causes an equivalent offset at the input:

$$v_{\text{off}(I_{\text{bias}})} = \frac{C_T}{C_{\text{in}}} \frac{\left[\sqrt{2I^+_{\text{bias}}} - \sqrt{2I^-_{\text{bias}}}\right]}{\sqrt{\beta_n}} \qquad (3.46)$$

This term, however, is much smaller than the previously discussed sources of offset and can be controlled by making the PMOS transistors sufficiently large.

In order to illustrate the performance of this cell, the comparator is designed in a 0.35 μm process with threshold voltages around 0.55 V. It operates at 0.9 V supply voltage. The power consumption measured at 11 MHz is under 6.5 μW, which compares favourably with other LP comparator realisations. The equivalent offset at the input, determined with Monte Carlo simulations is 5 mV with 3σ standard deviations of device variation. The operation is shown in Fig. 3.15.

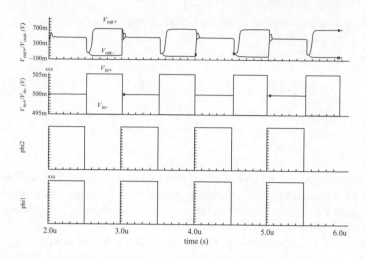

Figure 3.15   *Comparator transient simulation*

3.3.5   FGMOS D/A converters

This section illustrates how to use the FGMOS transistors to realise digital to analog conversion. The circuits presented here offer innovative techniques for facilitating D/A conversion by simplifying the topologies and hence reducing the required power supply voltage. It is left to the reader to further develop these ideas and produce novel LP/LV D/A converters.

3.3.5.1   Converter design 1

Figure 3.16 shows the schematic of a very simple D/A converter based on the operation of the FGMOS transistor M1. M1 has $N+2$ inputs, where N is the number of bits. N of the inputs are binary weighted. The other two inputs are weighted accordingly to keep transistors biased in the right operating region. The converter works as follows: if it is assumed that both n-channel devices have the same channel size, the p-channel transistors have the same dimensions and all four of them are operating in the strong inversion saturation region, then, using eq. (2.8)

$$I_{M1} = \frac{\beta_n}{2} \left(\frac{C_{in}}{C_T} \sum_{n=1}^{N} 2^{(n-1)} V_i + \frac{C_c}{C_T} V_{DD} - V_{Tn} \right)^2 \tag{3.47}$$

$$I_{M4} = \frac{\beta_n}{2} (V_{out} - V_{Tn})^2 \tag{3.48}$$

where I_{M1} and I_{M4} are the currents flowing through transistors M1 and M4 respectively, C_{in} is the minimum value of the binary weighted input capacitances and C_c is a biasing capacitance connected to $V_c = V_{DD}$. As a first approximation the parasitics have been neglected from the equation for I_{M1}. Transistor M2 mirrors the current I_{M1} to transistor M3, which makes $I_{M1} = I_{M4}$. Hence, the voltage generated at the

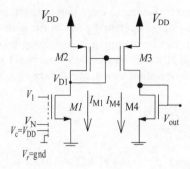

Figure 3.16   N-bit D/A converter based on the FGMOS

output is

$$V_{\text{out}} = \frac{C_{\text{in}}}{C_T} \sum_{n=1}^{N} 2^{(n-1)} V_i + \frac{C_c}{C_T} V_{\text{DD}} \tag{3.49}$$

Equation (3.49) is the function of a D/A converter if the inputs V_i are digital values [134].

Let us now explain the need and function of the biasing capacitances C_c and C_r[11]. The previous derivation has assumed that the four transistors are biased in the strong inversion saturation region. In order to achieve that, the following boundary conditions have to be maintained for transistors M1 and M4 (see Table 2.1):

Transistor M1

$$\frac{C_c}{C_T} V_{\text{DD}} > V_{\text{Tn}} \tag{3.50}$$

$$V_{\text{D1}} > \left(\frac{C_{\text{in}}}{C_T} \sum_{n=1}^{N} 2^{(n-1)} + \frac{C_c}{C_T} \right) V_{\text{DD}} - V_{\text{Tn}} \tag{3.51}$$

Transistor M4

$$V_{\text{out}} > V_{\text{Tn}} \tag{3.52}$$

Also, for the p-channel transistors to operate in the strong inversion saturation region:

Transistor M2

$$V_{\text{DD}} - V_{\text{D1}} > |V_{\text{Tp}}| \tag{3.53}$$

Transistor M3

$$V_{\text{out}} < V_{\text{D1}} + |V_{\text{Tp}}| \tag{3.54}$$

---

[11] C_r is the value of the input capacitance connected to V_r = ground in Fig. 3.16.

Equation (3.50) shows that by having an extra input capacitance C_c connected to V_{DD} it is possible to shift the voltage operating point at the gate of M1 by $(C_c/C_T)V_{DD}$. In this way, the transistor can be operating in the strong inversion saturation region even with the all zero digital word connected to the input. However, from eq. (3.54), in order to keep M3 in the saturation region, the voltage V_{out} (eq. (3.49)) has to remain below

$$V_{out} < \frac{V_{DD} + (\sqrt{\beta_n/\beta_p})V_{Tn}}{(1 + \sqrt{\beta_n/\beta_p})} \tag{3.55}$$

where β_p is the β parameter of transistors M2 and M3.

Hence, for the highest digital word, or what is the same, the most restrictive condition

$$\frac{C_r}{C_T} > \frac{\sqrt{\beta_n/\beta_p}}{(1 + \sqrt{\beta_n/\beta_p})} \left(1 - \frac{V_{Tn}}{V_{DD}}\right) \tag{3.56}$$

where as the first approximation the effect of the parasitics has been neglected thus yielding a value for the total capacitance

$$C_T = C_r + C_c + C_{in} \sum_{n=1}^{N} 2^{(n-1)} \tag{3.57}$$

Equation (3.56) shows how in order to keep transistor M3 saturated for all input combinations C_r has to be different from zero. Rewriting now eq. (3.50), the boundary conditions for C_r and C_c can be established:

$$\frac{\sqrt{\beta_n/\beta_p}}{(1 + \sqrt{\beta_n/\beta_p})} \left(1 - \frac{V_{Tn}}{V_{DD}}\right) < \frac{C_r}{C_T} < 1 - \frac{C_{in}}{C_T} \sum_{n=1}^{N} 2^{(n-1)} - \frac{V_{Tn}}{V_{DD}} \tag{3.58}$$

$$\frac{V_{Tn}}{V_{DD}} < \frac{C_c}{C_T} < 1 - \frac{C_{in}}{C_T} \sum_{n=1}^{N} 2^{(n-1)} - \frac{\sqrt{\beta_n/\beta_p}}{(1 + \sqrt{\beta_n/\beta_p})} \left(1 - \frac{V_{Tn}}{V_{DD}}\right) \tag{3.59}$$

Hence, for example, for $\beta_n/\beta_p = 0.1$, if the n-channel transistor threshold voltage is 0.5 V, and $V_{DD} = 1$ V, the required values of the compensating capacitances, normalised to the total capacitance, would be

$$0.045 < \frac{C_r}{C_T} < 0.5 - \frac{C_{in}}{C_T} \sum_{n=1}^{N} 2^{(n-1)} \tag{3.60}$$

$$0.5 < \frac{C_c}{C_T} < 0.9545 - \frac{C_{in}}{C_T} \sum_{n=1}^{N} 2^{(n-1)} \tag{3.61}$$

In order to maximise the step size for a given number of bits, the ratio C_{in}/C_T should be maximum, and hence, the minimum values of the compensating capacitances that fulfil the boundary conditions should be chosen. If $C_c/C_T = 0.5$, and for example

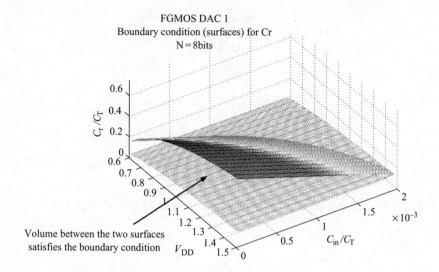

Figure 3.17   Boundary condition for C_r in the 8-bit DAC (Fig. 3.16)

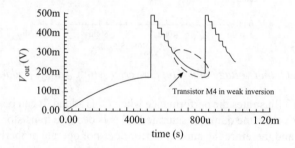

Figure 3.18   Output voltage of a 4-bit converter (Fig. 3.16)

$(C_{in}/C_T) \sum_{n=1}^{N} 2^{(n-1)} = 0.45$, then for $0.045 < C_r/C_T < 0.05$ both equations (3.60) and (3.61) would be met.

Figure 3.17 illustrates the design tradeoff represented by eq. (3.58). A value of C_r/C_T that meets all the design constraints has to be contained between the two surfaces. Once that value is chosen, C_c/C_T can be found from eq. (3.57).

An example of the design of this circuit is shown in the following. Figure 3.18 shows the output of a 4-bit converter designed using a 0.35 μm technology with $V_{DD} = 1$ V, when the input changes in time from the maximum to the minimum digital word (starting at time = 500 μs). It can be seen how for low values of the digital words the circuit does not have enough time to realise the conversion properly and the output appears distorted. This happens because for low values of the input, the output transistor M4 is in weak inversion.

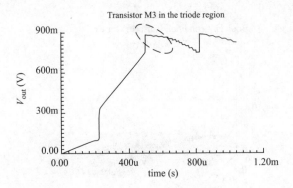

Figure 3.19   Output voltage of a 4-bit converter with C_c added (Fig. 3.16)

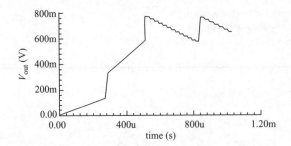

Figure 3.20   Output voltage of a 4-bit converter with C_c and C_r added (Fig. 3.16)

Figure 3.19 illustrates the performance when C_c is added. It can be observed how for high input values the output is saturated. This is because transistor M3 enters the triode region and therefore the current mirror does not operate properly anymore.

Finally, the operation of the same converter with the two biasing capacitances is shown in Fig. 3.20. The figure shows how the converter is linear for the whole operating range.

Examples of the operation of a 6-bit and an 8-bit converter are also shown in Fig. 3.21 and Fig. 3.22. The performance of these blocks is summarised in Table 3.8. The Integral Non Linearity (INL) and Differential Non Linearity (DNL) of the 6-bit A/D converter is illustrated in Fig. 3.23.

Also, Monte Carlo simulations were run for both designs. They showed that the main problem this topology might have would be in the form of an output offset. This could be compensated just by changing the value of the bias voltage connected to C_c. Another possibility would be to design some kind of self-compensation strategy, similar to the one used for the comparator described in the previous section.

These designs are not optimised and could be improved by, for example, using cascode transistors. They would increase the output resistances of the devices, and this would allow the use of a smaller C_{in} capacitance in the FGMOS device and would improve the speed and linearity. The disadvantage would be a more limited output range.

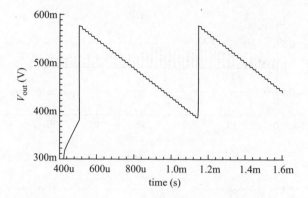

Figure 3.21   Output voltage of a 6-bit converter (Fig. 3.16)

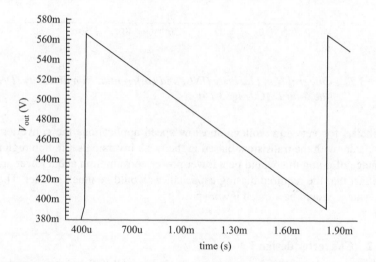

Figure 3.22   Output voltage of an 8-bit converter (Fig. 3.16)

Table 3.8   Summary of performance for the converter in Fig. 3.16

	6-bit	8-bit
LSB	3 mV	0.75 mV
Non linearity and noise	<1LSB	<1LSB
Maximum sampling rate	11 MS/s	11 MS/s
Average power consumption	0.5 μW	0.5 μW

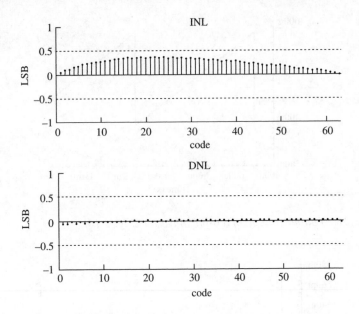

Figure 3.23   Integral Non Linearity (INL) and Differential Non Linearity (DNL) for the 6-bit DAC (Fig. 3.16)

Finally, for very low voltage and low speed applications the converter could also operate with the transistors biased in the weak inversion saturation region. The advantage of doing this would be a lower power consumption and a lower area due to the fact that the required biasing capacitances would be much smaller. The main disadvantage would be a much lower speed.

3.3.5.2   Converter design 2

Another example of D/A converter based on the FGMOS transistor is shown in Fig. 3.24. In this topology, the circuit operation is based on the transistor M1 working in the strong inversion ohmic region. Again, a biasing capacitor, C_c, connected to V_{DD} is used to bias the input transistor in the strong inversion region. The other input capacitances are binary weighted. The opamp fixes the voltage at the drain of M1 to V_{ref}. Hence, the current flowing through M1 and M2 is given by

$$I_{M1} = \beta_n \frac{C_{in}}{C_T} V_{ref} \sum_{n=1}^{N} 2^{(n-1)} V_i + \beta_n \left(\frac{C_c}{C_T} V_{DD} - V_{Tn} \right) V_{ref} \qquad (3.62)$$

And again if the inputs are digital voltages, this represents the function of a D/A converter. This current is mirrored to an output branch where it can be converted back to a voltage just by using a resistance. The latter can be implemented using active components, although, in this example, a passive resistor has been considered

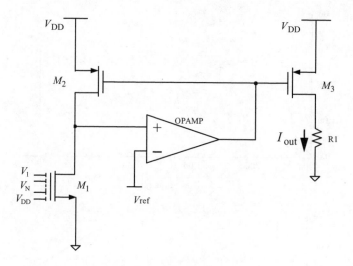

Figure 3.24   N-bit D/A converter based on an FGMOS transistor operating in the ohmic region

for the sake of simplicity. The second term in eq. (3.62) represents an offset at the output. The offset can be cancelled out by using, for example, a differential topology. Another way to compensate for it is by using a tuning technique which either adds a current at the output, or connects the biasing capacitor to a variable bias voltage.

The performance of 6-bit and 8-bit D/A converters (as in Fig. 3.24) is shown in Fig. 3.25. The average power consumption of this converter is around 10 μW. It is higher than the power consumption of the converter described in the previous section due to the additional power required by the opamp. Also, the voltage supply used is 1.5 V as opposed to 1 V in the previous topology. For both topologies the total capacitance is around 7 pF. This converter is again designed in a 0.35 μm technology with a value of V_{Tn} for the n-channel devices of around 0.5 V. The maximum sampling rate for the converter in Fig. 3.24 is 2 Ms/s.

3.3.6   A programmable switched-current floating gate MOS cell

This section explains how to use the FGMOS transistor in a switched-current (SI) memory cell as a means to increase its programmability capability.

SI circuits perform discrete time processing with analog circuits. Applications include filters and various types of ADC and DAC converters. The constituent parts of these circuits, SI cells, may be designed to have a gain, α, such that the z-domain transfer function of the cell is

$$i_{out} = \alpha \times z^{-1/2} \times i_{in} \qquad (3.63)$$

The gains of SI cells determine the overall transfer function of an SI circuit. Generally, the gain coefficients within SI cells are created by using current mirrors in various

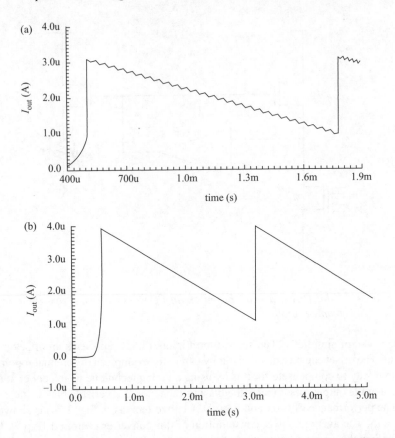

Figure 3.25   Output current for the converter in Fig. 3.24: (a) 6 bits (b) 8 bits

forms. In a first generation cell, Fig. 3.26(a), the current mirror is part of the cell. In a second generation cell, a single transistor is used for both input and output currents so a current mirror after the output is required to provide the necessary gain, as shown in Fig. 3.26(b) [135]. It is often useful, however, to adjust coefficients within switched current circuits after manufacture, to program them. This allows the circuit characteristics to change over time, for example, to provide adaptive filtering or to tune the circuit accounting for device mismatch and process variation.

A number of possible programming schemes are in the literature [136–142]. Programmability may be achieved with an array of different gain circuits that are switched between [136–138]. A discrete number of gains are possible by switching in and out parts of the array. This technique, however suffers from the limited size of the array. To increase the number of discrete values the gain may take, or resolution, requires increasingly more circuitry. Therefore this technique is only useful for a considerable range of gain and continuous programmability. A four quadrant multiplier built into a current cell [139] and a separate multiplier [140] after the SI cell are both found in the literature. However a multiplier is required for every SI cell

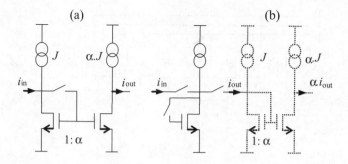

Figure 3.26   First generation SI cell (a) and second generation cell (b)

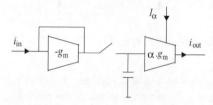

Figure 3.27   A first generation switched transconductance cell

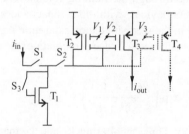

Figure 3.28   The PSIFG cell

thus requiring considerable circuitry for each cell. Finally, transconductance amplifiers can be used to program a circuit [141,142]. An example is the first generation
switched transconductance cell, Fig. 3.27, where the gain, α, is varied by current I_α.
This solution requires considerable circuitry, with an amplifier for each cell, and, the
programmability is of limited range. All these techniques either offer a limited resolution of gain or require considerable circuitry per current cell for programmability.
In low power applications, where programmability is necessary, the large amounts
of circuitry consume substantial power. This is undesirable. A simple way to add
programmability without increasing the circuitry and hence reducing the power is
by using the FGMOS transistor in a configuration that from now on will be called
Programmable Switched Current Floating Gate Cell (PSIFG).

Figure 3.28 shows the basic PSIFG cell. T_1 provides the sample and hold element
necessary for SI cells while T_2 and T_3 form a floating gate current mirror whose

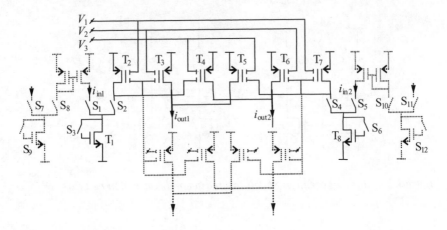

Figure 3.29   The differential PSIFG cell

current gain is programmed by the bias voltages, V_1 and V_2. Switches S_1 and S_3 are closed and S_2 open during the sampling phase of the clock signal. The voltage on the gate of T_1 is held during the second phase when S_1 and S_3 are opened and S_1 closed. This clocking strategy is similar to a second generation SI cell. During the hold phase, T_1 drain current is steered through the floating gate current mirror. Assuming two-input (V_1 and V_2) FGMOS devices, with input capacitances C_1 and C_2 (for V_1 and V_2, respectively), biased in the weak inversion saturation region, applying eqs. (2.5) and (3.63), neglecting parasitics as a first approximation, the overall transfer function for the circuit in Fig. 3.28 is

$$i_{out} = z^{-1/2} \times i_{in} \times \exp\left(\frac{k(V_2 - V_1)}{\eta U_T}\right) \tag{3.64}$$

where $k = C_2/(C_2 + C_1)$. Thus, by altering V_1 and V_2 the gain of the cell may be changed.

The basic PSIFG cell may be extended to a differential version, solid lines of Fig. 3.29. The cell is formed from two basic PSIFC cells: T_1, T_2 and T_3 form one cell and T_6, T_7 and T_8 form another. Transistors T_4 and T_5 extend the respective floating gate current mirrors giving the differential PSIFG cell its principle advantage: the gain may be both positive and negative, unlike the simple PSIFG cell with only positive gain. This may be observed in the mathematics, where the expressions for the output currents, i_{out1} and i_{out2}, are

$$i_{out1} = z^{-1/2} \times i_{in1} \times \exp\left(\frac{k(V_2 - V_1)}{nU_T}\right) + z^{1/2} \times i_{in2} \cdot \exp\left(\frac{k(V_3 - V_1)}{nU_T}\right)$$

$$i_{out2} = z^{-1/2} \times i_{in1} \times \exp\left(\frac{k(V_3 - V_1)}{nU_T}\right) + z^{1/2} \times i_{in2} \cdot \exp\left(\frac{k(V_2 - V_1)}{nU_T}\right)$$

$$\tag{3.65}$$

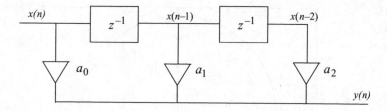

Figure 3.30   Block diagram of a second-order FIR filter

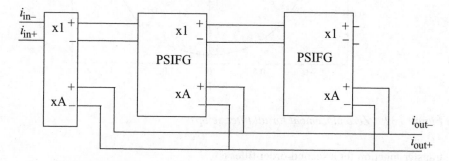

Figure 3.31   Realisation of the filter from PSIFG cells

Hence the differential output, i_{out}, is given by

$$i_{out} = i_{out1} - i_{out2} = z^{-1/2} \times (i_{in1} - i_{in2}) \times \left(\exp \left(\frac{k(V_2 - V_1)}{nU_T} \right) \right.$$

$$\left. - \exp \left(\frac{k(V_3 - V_1)}{nU_T} \right) \right) \tag{3.66}$$

Thus, when $V_2 > V_3$ the cell gain is positive and when $V_2 < V_3$ it is negative. Also, two extensions can be added to the differential PSIFG cell as shown by the dotted sections of Fig. 3.29: additional outputs may be produced by extending the floating gate current mirror; and an extra switched circuit may be added at the inputs to create a full unit delay, z^{-1} from the $z^{-1/2}$ already present. This solution for programmable SI cells is advantageous as it does not require any form of transconductance or operational amplifier for each cell, reducing power consumption and it provides a continuous gain factor unlike array switching.

The concept of PSIFG is illustrated in the following with the design of a second order FIR filter. The equation for the filter is

$$y(n) = a_0 x(n) + a_1 x(n-1) + a_2 x(n-2) \tag{3.67}$$

This may be seen in Fig. 3.30 in the form of a block diagram. Two PSIFG cells are required, both with a full unit delay and one with a double set of outputs, Fig. 3.31. In addition to the SI cells, a floating gate current mirror is used at the input of the system to provide the coefficient a_0, without any delay added. In the z domain the

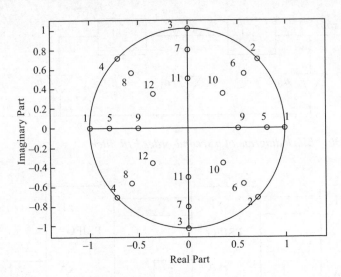

Figure 3.32   Zero placement for all filter setups

transfer function for a second-order filter is

$$T(z) = \frac{Y(z)}{X(z)} = a_0 + a_1 z^{-1} + a_2 z^{-2} \qquad (3.68)$$

Thus, for a_2, a_1, a_0 to be real, the roots of $T(z)$ are either a complex conjugate pair or both real, the roots of $T(z)$ being the zeros for the system. For a pair of complex conjugate roots of radii r and at angles $\pm\varphi$, the coefficients are

$$a_1 = -2a_0 r \cos(\varphi) \qquad (3.69)$$

$$a_2 = a_0 r^2 \qquad (3.70)$$

where a_0 is chosen as a suitable scaling factor given the range of gains from the floating gate current mirrors. The two poles of the system are located at the origin making the filter inherently stable.

To illustrate the filter performance, a range of values for r and φ are chosen, $r = 0.5$, 0.8 and 1, $\varphi = 45$, 90, 135 degrees. In addition three points for the real-valued zeros on the real axis are chosen, with zero pairs at 1, -1; 0.8, -0.8; and 0.5, -0.5. All the zero positions may be seen in Fig. 3.32. The numbers correspond to each pair of zeros, for identification.

With a_0 set to 0.5, Table 3.9 shows the values of a_1 and a_2 where the zeros position number corresponds to the numbers in Fig. 3.32.

The filter is designed in a $0.35\ \mu$m CMOS process and simulated using a 1 V power supply. The clock frequency is 33.3 kHz. Figure 3.33 is a timing diagram for the various clocks signals. Switches controlled by φ_1 and φ_4, S_1, S_2, S_4, S_5, S_7, S_8, S_{10} and S_{11} of Fig. 3.29, are single NMOS FETs. The other switches have dummy transistors included, of half the width of the switching transistor. Dummy switches

Table 3.9   Values of the coefficients corresponding to the zeros in Fig. 3.32

Zero position number	Radius (r)	φ	a_1	a_2
1	1	0 (and 180)	0	0.5
2	1	45	−0.707	0.5
3	1	90	0	0.5
4	1	135	0.707	0.5
5	0.8	0 (and 180)	0	0.32
6	0.8	45	−0.566	0.32
7	0.8	90	0	0.32
8	0.8	135	0.566	0.32
9	0.5	0 (and 180)	0	0.125
10	0.5	45	−0.354	0.125
11	0.5	90	0	0.125
12	0.5	135	0.354	0.125

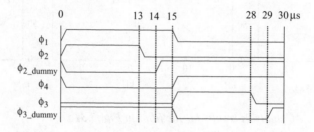

Figure 3.33   Timing diagram for the simulations

switch slightly after the main switch, so as to not interfere with its operation. Signal φ_2 controls switches S_9 and S_{12}, signal φ_3 controls switches S_3 and S_6.

Transfer functions for the FIR filter with the various coefficients are shown in Fig. 3.34. The figure shows transfer functions for the values of coefficients chosen. The x axis is frequency in kHz and the y axis is amplitude in nA. The ideal response is dotted, the simulated response is a continuous line. The numbers in each plot correspond to the numbers of each pair of zeros on the z domain plot, Fig. 3.32. Columns show the zeros with radii, 1, 0.8 and 0.5 from top to bottom. Rows show the zeros as they move around the unit circle, 0 (and 180), 45, 90 and 135 degrees from left to right. Other characterisation results are shown in Table 3.10.

3.4   Summary and conclusions

The FGMOS transistor can be seen either as a MOS transistor with an analog adder circuit connected at its gate, or as an MOS transistor with an electrically tunable threshold voltage, or even as a complex nonlinear multiplier divider block. Besides,

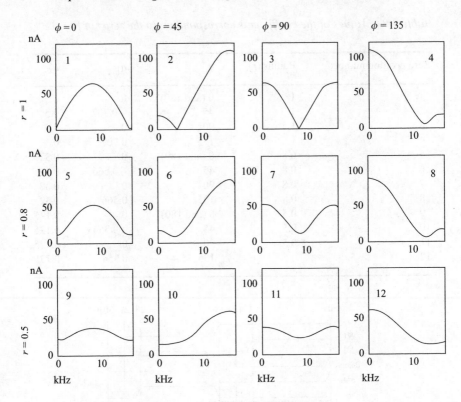

Figure 3.34   *AC analysis results for the filter in Fig. 3.31*

Table 3.10   *Summary of performance
for the filter in Fig. 3.31*

Modulation index	0.6
Input amplitude	30 nA
THD	Under 0.5%
Supply voltage	1 V
Power consumption	1 μW
Clock frequency	33.3 kHz

the degrees of freedom the designer now has to play with increases from 3 to $N+2$ for a single device, where N is the number of inputs in the FGMOS. This opens up a new range of possibilities for the designer in terms of design tradeoffs. Although this is very promising, designing becomes even more complex. This chapter has illustrated with simple examples how to obtain the benefits of the FGMOS. In addition, an overview has been given on how to foresee when the device might or might not be useful in different contexts, all of them having in common the low power constraint.

Notation

α	Gain coefficient in a SI memory cell (Fig. 3.27)
β_i	Mi β parameter
β_{FG}	FGMOS equivalent β parameter (eq. (3.2))
β_n	β parameter of n-channel MOS device
β_p	β parameter of p-channel MOS device
$\Delta U_{inverter}$	Variation of $\Delta U_{inverter}$ caused by threshold voltage variations (eq. (3.26))
ΔV_{DD}	Difference between the minimum voltage supply needed by two current mirror cells, a conventional one and the FGMOS based cell (Fig. 3.3, eq. (3.14))
$\Delta V_{off}^{+}, \Delta V_{off}^{-}$	Variation of the voltages at nodes V_{off}^{+}, V_{off}^{-} during the comparison phase with respect to the values they have at the end of compensation phase in the FGMOS comparator (Fig. 3.13)
ΔV_T	$V_{Tn2} - V_{Tn1}$ difference between the threshold voltages of M1 and M2 in the FGMOS comparator (Fig. 3.12, eq. (3.32))
ΔV_{Tn}	Threshold voltage variation between n-channel devices in Fig. 3.10 (eq. (3.26))
$\Delta\lvert V_{Tp}\rvert$	Threshold voltage variation between p-channel devices in Fig. 3.10 (eq. (3.26))
φ	Angles of complex conjugate roots of T(z)
A_o	Total DC gain of the two-stage MOS opamp in Section 3.3.2
A_{oFG}	Total DC gain of the two-stage FGMOS opamp in Section 3.3.2
a_i	Coefficients of the filter in eq. (3.67)
C_1	Input capacitance connected to V_I in the FGMOS current mirror cell (Fig. 3.3). Also input capacitances in several of the FGMOS transistors in the opamps in Fig. 3.8 and Fig. 3.9 (their values are in Table 3.3 and Table 3.6).
$(C_1/C_T)_i$	Mi input weight (eq. (3.22), Fig. 3.9)
C_2	Input capacitance used for biasing purposes in several FGMOS devices in the opamps in Fig. 3.8 and Fig. 3.9 (their values are in Table 3.3 and Table 3.6).
C_c	Value of the capacitance connected to V_c in Fig. 3.10. Also, value of the capacitance connected to the offset compensation signal in the comparator in Fig. 3.12 and of the biasing capacitance connected to $V_c = V_{DD}$ in the N-bit D/A converters in Fig. 3.16 and Fig. 3.24.
C_{clk}	Value of the capacitance connected to the clock signal in the FGMOS comparator in Fig. 3.12
C_{comp}	Compensating capacitance for the opamps in Section 3.3.2
C_f	Small extra input capacitance added to compensate the offset due to C_{GD} and feedthrough in the FGMOS comparator in Fig. 3.13 (eq. (3.45))
C_{GB}	Gate-to-bulk capacitance

C_{GBp}	Gate-to-bulk capacitance of p-channel FGMOS device in Fig. 3.10. (Eq. (3.25). Used when it is necessary to differentiate between n-channel and p-channel devices.)
C_{GD}	Gate-to-drain parasitic capacitance
C_{GDn}	Gate-to-drain capacitance of n-channel FGMOS device in Fig. 3.10. (Eq. (3.25). Used when it is necessary to differentiate between n-channel and p-channel devices.)
C_{GDp}	Gate-to-drain capacitance of p-channel FGMOS device in Fig. 3.10. (Eq. (3.25). Used when it is necessary to differentiate between n-channel and p-channel devices.)
$(C_{GD}/C_T)_i$	Ratio between gate-to-drain capacitance and the total capacitance for transistor Mi (eq. (3.23))
C_{GSp}	Gate-to-source capacitance of p-channel FGMOS device in Fig. 3.10. (Eq. (3.25). Used when it is necessary to differentiate between n-channel and p-channel devices.)
C_i for $i = [1, N]$	FGMOS input capacitances
C_{in}	Input capacitance connected to the effective input of the FGMOS comparator in Fig. 3.12 and also in the inverter in Fig. 3.10 (used when there is only one effective input) (In the case of the N-bit D/A converter in Fig. 3.16 it is also the minimum value binary weighted input capacitance)
C_r	Input capacitance connected to $V_r =$ ground in the N-bit D/A converter in Fig. 3.16
C_T	Total capacitance seen by the FG (the same value will be assumed for all the devices unless the opposite is said)
C_{VDD}	Input capacitance connected to V_{DD}
GBW	Gain-bandwidth product of a two-stage MOS opamp
GBW_{FG}	Gain-bandwidth product of a two-stage FGMOS opamp
g_{dsFi}	Equivalent output conductance for the FGMOS transistor Mi
g_{dsi}	Transistor Mi output conductance
g_{mi}	Transistor Mi gate transconductance
I_α	Current controlling the gain α in switched transconductance cell (Fig. 3.27)
I_{bias}	Current flowing through PMOS in the comparator transistors M3 and M4 in the absence of mismatch between them when V_{bias} is applied (Fig. 3.13)
I_{bias}^-, I_{bias}^+	Different bias currents flowing through the negative and positive comparator (Fig. 3.13) branches, respectively, due to the mismatch between the threshold voltages in top PMOS transistors during the compensation phase
I_{in}^-	Current in the input branch of the current mirror in Fig. 3.3
I_{in}^{max}	Maximum value of input current I_{in} in the current mirror in Fig. 3.3 (eqs. (3.11)–(3.14))

I_j for $j = [1, N]$	Current flowing through j-th MOS transistor in Fig. 3.2 (eq. (3.7))
I_{Mi} for $i = [1, 4]$	Current flowing through Mi in the N-bit D/A converters in Fig. 3.16 and Fig. 3.24
I_{out}	Output current
I'_o	See Fig. 3.2
i_{in}	Input current in the SI cells in Fig. 3.26 and Fig. 3.28
i_{in-}	Single input current for the differential filter built with PSIFG cells (Fig. 3.31)
i_{in+}	Single input current for the differential filter built with PSIFG cells (Fig. 3.31)
i_{ini} for $i = [1, 2]$	Input currents in the differential PSIFG cell in Fig. 3.29
i_{out-}	Single output current for the differential filter built with PSIFG cells (Fig. 3.31)
i_{out+}	Single output current for the differential filter built with PSIFG cells (Fig. 3.31)
i_{outi} for $i = [1, 2]$	Output currents in the differential PSIFG cell in Fig. 3.29
J	Bias current source in Fig. 3.26
k	$C_2/(C_2 + C_1)$ (eqs. (3.64)–(3.66))
L_i	Effective length of transistor Mi
phi1	Clock signals for S_1 and S_2 in FGMOS comparator in Fig. 3.13
phi2	Clock signals for S_3 and S_4 in FGMOS comparator in Fig. 3.13
phi2_b	Clock signals for S_5 and S_6 in FGMOS comparator in Fig. 3.13
Q_{FG}	Charge trapped at the FG
R_c	Compensating resistance in the two-stage opamps
r	Radius of complex conjugate roots of T(z) (zeros of the system) (eqs. (3.69) and (3.70), Table 3.9)
$T(z)$	Transfer function for the second-order filter in eq. (3.68)
$U_{inverter}$	Switching threshold of CMOS inverter (eq. (3.24))
V_1	M1 drain-to-source voltage in Fig. 3.3 (eqs. (3.9)–(3.11)). Also one of the voltages used to tune the PSIFG cell (Fig. 3.28)
V_2	One of the voltages used to tune the PSIFG cell (Fig. 3.28)
V_{bias}	Bias voltage of PMOS transistors M3 and M4 in FGMOS comparator in Fig. 3.13
V_{bias1}	Bias voltage for M1 in the current mirror with FGMOS in Fig. 3.3(b) (assumed to be equal to V_{DD})
V_{bias2}	Bias voltage for M3 in the current mirror with FGMOS in Fig. 3.3(b) (assumed to be equal to V_{DD})
V_c	Input voltage of the two-input FGMOS inverter in Fig. 3.10 – used for changing the switching threshold (eq. (3.25))
V_{clk}	Comparator clock voltage
V_{cm}	Common mode voltage applied to V_{in}^+ during the offset compensation phase (Fig. 3.13)

V_{D1}	Drain voltage of transistor M1 in N-bit D/A converter in Fig. 3.16
V_{D3}	Drain-to-source voltage of transistor M3 in Fig. 3.3(b)
V_{DD1}	Voltage at the drain (gate) of diode connected transistor M2 in the cascode current mirror in Fig. 3.3 (eq. (3.10))
V_{DD}^{min}	Minimum voltage supply needed by the current mirror in Fig. 3.3 to drive a maximum input current I_{in}^{max} (eqs. (3.11)–(3.14))
V_{FG}	Voltage at the FG (Fig. 3.1)
V_i for $i = [1, N]$	FGMOS effective inputs
V_{in}	FGMOS effective input voltage (when there is only one)
$V_{ind}^{(effective)}$	Effective differential input signal when there is an offset caused by mismatch between the input transistors (Fig. 3.13, eq. (3.39))
$V_{in(threshold)}$	Voltage at the effective input of FGMOS inverter for which the switching of the inverter happens
$V_{in}^{\pm}$	Effective inputs of the FGMOS comparator in Fig. 3.12
V_{Load}	Voltage across the biasing/input part of the cascode mirror circuit in Fig. 3.3
V_{off}	Input of the FGMOS comparator in Fig. 3.12 used for offset compensation. It is $V_{off}^+ = V_{off}^-$ when there is no mismatch
V_{off}^+, V_{off}^-	Voltage generated at V_{out}^+ and V_{out}^- respectively (by the negative feedback) in the FGMOS comparator in Fig. 3.13 during the offset compensation phase (eqs. (3.34) and (3.35))
$V_{off(C_{GD}/feedthrough)}$	Equivalent input referred offset due to the effect of the parasitic capacitance C_{GD} and feedthrough (eq. (3.45))
$V_{off(I_{bias})}$	Equivalent offset at the input of the comparator in Fig. 3.13 due to the difference in bias currents in the positive and negative branches of the comparator (eq. (3.46))
$V_{off \Delta V_T}$	Offset at the input caused by threshold voltage variations in the FGMOS comparator (Fig. 3.12, eq. (3.32))
$V_{out}^{\pm}$	Single outputs of the FGMOS comparator in Fig. 3.13
$V_{out(comparison)}^-,$ $V_{out(comparison)}^+$	Voltage at the '$-$' or '$+$' output node of FGMOS comparator in Fig. 3.13 during the comparison phase
$V_{out(compensation)}^-,$ $V_{out(compensation)}^+$	Voltage at the '$-$' or '$+$' output node of FGMOS comparator in Fig. 3.13 during the compensation phase
V_{out1}	Output of the FGMOS inverter in Fig. 3.10
V_{ref}	Voltage at the drain of M1 in the D/A converter in Fig. 3.24 fixed by the opamp (eq. (3.62))
V_T'	$(C_T/C_i)V_{TFG}$. FGMOS effective threshold voltage parameter (eq. (3.3))

V_{TFG}	$V_{\text{T}} - \sum_{\substack{j=1 \\ j \neq 1}}^{N} (C_j/C_{\text{T}})V_{j\text{S}} - Q_{\text{FG}}/C_{\text{T}}$. Different value for different designs within the chapter
$V_{\text{Tn}i}$ for $i = [1, 2]$	Mi threshold voltage when there is mismatch in the FGMOS comparator (Fig. 3.12)
V_{Tn}	n-Channel MOS transistors threshold voltage (the same for all the devices unless the opposite is said)
$|V_{\text{Tp}}|$	p-Channel MOS transistors threshold voltage (the same for all the devices unless the opposite is said)
W_i	Effective width of transistor Mi
w_j	C_j/C_{T}. Input weight corresponding to the j-th FGMOS input
$w_{\text{p}i}$ for $i = [1, 2]$	Angular frequency of the i-th pole in the MOS/FGMOS two stage opamp (eq. (3.21))
$x(n)$	See eq. (3.67) (FIR filter)
$y(n)$	See eq. (3.67) (FIR filter)

Chapter 4

Low power analog continuous-time filtering based on the FGMOS in the strong inversion ohmic region

4.1 Introduction

The following four chapters describe how to exploit the mathematical capabilities of FGMOS devices in more complex circuits taking as an example the design of continuous-time analog filters. Filters are inevitable building blocks in high-performance low power (LP) and low voltage (LV) electronics [143,144]. A filter circuit is a two-port unit and its function is to process the magnitude and/or phase of an input signal in a desired way. There are several filter realisations and their choice depends on the needed signal processing as well as the required circuit performance [145]. Once the mathematical function describing a filter transfer characteristic is derived, its final realisation will be determined by technological limitations.

The first decision to be made when opting for a filter topology is whether it is going to be a discrete or a continuous-time system [146]. In many cases, the filter input signal comes from the real world and therefore is continuous in time. At this point, the designer has to evaluate whether it is beneficial for the whole system performance to implement it all in discrete time or not. The best performing technique for designing discrete time filters is the one known as switched capacitor. Switched capacitor circuits simulate filters transfer functions only in terms of capacitances ratios and so component tolerancing is not a major problem. However, switching speed limitations make this a fairly low frequency technique.

Yet, the main focus of this book is on continuous-time circuits[12]. There are three main approaches for the design of CMOS continuous-time active filters: MOSFET-C filters, RC active filters and OTA-based filters [147].

---

[12] Some discrete time realisations can be found in Chapter 3.

MOSFET-C filters are based on the use of an equivalent CMOS tunable resistor in an RC structure [148,149]. Their main drawback is that they require high-performance OPAMPs which limits their use to low-frequency applications. Besides, if the voltage supply is scaled down it becomes very difficult to keep an acceptable performance since the OPAMP functionality seriously degrades.

The difference between RC filters and MOSFET-C filters is that the former use passive instead of MOSFET-based resistors. This adds some extra problems as they require a technological process with linear, stable resistors which also sometimes need to be large and well matched. Furthermore, it is more difficult to tune them.

The basic elements of OTA-C filters[13] are Operational Transconductance Amplifiers (OTA) and capacitors (C). The major advantage of OTA-based continuous-time filters versus MOSFET-C realisations is their wider frequency response [150–159]. In MOSFET-C blocks the frequency range is limited to one-thirtieth of the OPAMP gain-bandwidth frequency (GBW), whereas in OTA-based integrators which are implemented in open loop configuration the only limitation to the frequency response is the OTA gain-bandwidth product [160]. The main drawback of OTA realisations is the smaller dynamic range which is limited by the distortion introduced by nonlinear terms in the mathematical functions that describe the current. Linearised topologies can be used in order to reduce higher order harmonic components [153,161–178]. Owing to all these reasons, linearised OTA-C topologies seem to appear as the most suitable choice when the predominant design constraints are the low voltage and low power [143].

The performance of low voltage continuous-time filters is directly related to the quality of the integrator building block measured in terms of power consumption, linearity, noise, phase response and tunability as main parameters [160]. There are different approaches in terms of the integration method, which can be either the traditional internally linear-externally linear (ILEL) or externally linear-internally nonlinear (ELIN) [21,22,179–188]. The latter has appeared as a very popular option for the low-voltage high-frequency operation in bipolar or BiCMOS technologies [189–193] while MOS realisations are more suitable for audio frequencies [26,194,195]. An important subset of ELIN circuits are logarithmic filters. Their operation is based on the exponential law that models the behaviour of devices such as bipolar transistors or MOS working in the weak inversion region [169,170,179,182,189–205].

The following four chapters describe a different approach in designing continuous-time analogue filters, which combines some of the previously reported techniques with the use of FGMOS devices within the building blocks. This design approach has proven to be very useful when re-establishing the technological limits in order to reduce the power [94,101,174,177,178,206–210]. The aim of these chapters is to teach the reader how to take advantage of the FGMOS device when the main design constraints are the power supply voltage and power consumption and to analyse the negative implications of using this device.

---

[13] Also called G_m-C filters.

The first chapter of this series of chapters focuses on the properties of an FGMOS transistor operating in the strong inversion ohmic region. An LP/LV multiple-input linear transconductor is designed and subsequently used for the implementation of two circuits: a fully differential G_m-C integrator and a filter.

The FGMOS transistor is also used, although not biased in the ohmic region, in other sections of the filter circuit with different aims. Examples are the use of FGMOS devices (instead of normal MOS) as: a) cascode devices with reduced effective threshold voltages; or b) in a common mode feedback circuit to perform averaging at a very low power cost.

The side effects of using the FGMOS in each individual case are also thoroughly analysed. Studies of Power Supply Rejection Ratio (PSRR), Common Mode Rejection Ratio (CMRR) and Total Harmonic Distortion (THD) are provided.

Experimental results supporting the conclusions are shown at the end of the chapter.

4.2   Transconductor basic blocks

This section introduces basic building blocks that are later used in the implementation of a G_m-C integrator and a filter. Two circuit topologies are presented, namely N1 and N2. N1 represents a multiple input transconductor which uses an FGMOS transistor biased in the strong inversion ohmic region while N2 is a single stage FGMOS-based OPAMP required to linearise the operation of N1. The operation of both blocks, N1 and N2, is explained below.

The current law for an n-channel FGMOS transistor biased in the strong inversion ohmic region, as given by eq. (2.6), can be rewritten in a simpler form if the drain to source voltage is assumed to be very small, that is

$$V_{DS} \ll (V_{GS} - V_{Tn}) \tag{4.1}$$

In this situation, the quadratic term can be neglected from the equation, resulting in

$$I_D \approx \beta_n \left[\left(\sum_{i=1}^{N} \frac{C_i}{C_T} \cdot V_i - V_{Tn} \right) V_D \right] \tag{4.2}$$

where the n subscript in the threshold voltage as well as in β parameter is used to refer to the n-channel nature of the transistor. It has also been assumed that the source and bulk are connected to ground, there is no charge accumulation and the parasitic capacitances are much smaller than the input ones. Equation (4.2) shows that if the drain voltage, V_D, is kept constant, the current I_D will depend linearly on the input voltages with slopes $\beta_n(C_i/C_T)$.

An FGMOS-based implementation of eq. (4.2) is shown in Fig. 4.1[14]. M1 is the FGMOS transistor biased in the strong inversion ohmic region. Transistor M2

---

[14] This basic cell was also the basis of a D/A converter as explained in Chapter 3, Section 3.3.5.

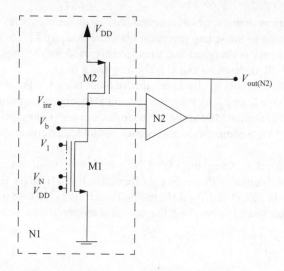

Figure 4.1   N1 – Basic circuit implementing eq. (4.2) (N1)

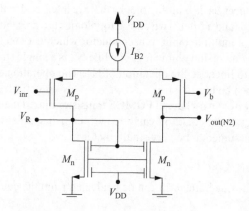

Figure 4.2   N2 – Tuning amplifier

performs a current to voltage conversion and will be used for the mirroring purposes. N2 is a high gain differential input/single output amplifier that sets the drain voltage of M1 to a constant reference: $V_D = V_b$. An FGMOS-based realisation of N2 is shown in Fig. 4.2. It consists of a pMOS differential pair with n-channel FGMOS transistors acting as a load. The input V_{inr} is compared with the reference voltage V_b and if their values are not equal the difference is amplified at the output $V_{out(N2)}$, which is connected to the gate of the pMOS transistor M2 (in N1 block). This would unbalance the currents flowing through M1 and M2 and would have to be corrected by the output resistance which changes the V_{inr} value in the opposite direction. The feedback forces the operating point of V_{inr} to be equal or almost equal to the voltage V_b. From now on this V_b voltage will be referred to as the tuning voltage.

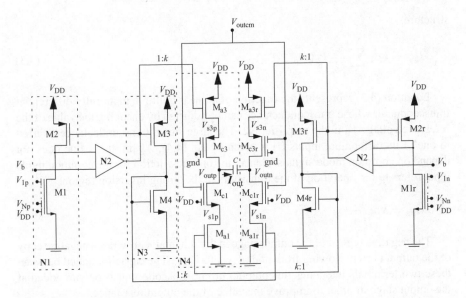

Figure 4.3 Fully differential multiple-input integrator[15]

4.3 The G_m-C multiple-input linear integrator. Large signal description

Figure 4.3 shows the schematic of a fully balanced linear transconductor built by replicating and combining the blocks described in the previous section.

The schematic shows two identical N1 blocks on the right and left side of the figure. The positive inputs are connected to the left side transconductor block (V_{jp}, $j = [1,N]$), whereas the negative ones are applied to the transconductor on the right side (V_{jn}, $j = [1,N]$). The respective output in each N1 cell is connected in a mirror configuration to a pMOS device (M3/M3r). The latter generates a copy of the current flowing through M1/M1r and sources that current to a diode connected nMOS transistor (M4/M4r) which, acting as the input device of a mirror structure, produces a negative/positive copy of it. A positive and a negative scaled (1 : k) replica of the negative and positive currents[16], respectively, are subtracted at the output nodes (V_{outp}/V_{outn}) generating the output current I_{out}, described by eq. (4.3). This equation shows how the constant term including the threshold voltage in eq. (4.2) disappears thanks to the differential nature of the

---

[15] The subscript r is used to distinguish between different devices with the same dimensions. Hence, for example, transistor M1 and transistor M1r will be equally sized. Also, from now on the parameters will have an added subscript in agreement with the transistor name.

[16] The positive current is the current generated by the p inputs while the negative current is the current generated by the n inputs.

structure:

$$I_{\text{out}} = k\beta_1 \sum_{j=1}^{N} \frac{C_j}{C_T} V_b (V_{jp} - V_{jn}) \qquad (4.3)$$

Equation (4.3) represents the function of a linear multiple input fully differential transconductor whose transconductance can be adjusted by changing the value of the reference voltage V_b in the amplifier N2. The large signal transconductance referred to one of its differential inputs (i) is given by eq. (4.4). Again, this value does not depend on common mode signals. Hence, the only restriction for the common mode input refers to the operation of the transistors in the strong inversion region:

$$G_{\text{mi}} = k\beta_1 \frac{C_i}{C_T} V_b \qquad (4.4)$$

The negative replica of the output stage in the schematic provides a negative copy of the output current flowing from node V_{outn} to V_{outp}. The capacitor added between these two terminals integrates this output current. Cascode transistors are added at the output stage in order to improve the value of the output resistance.

4.3.1   Operating point – design considerations

The biasing of certain devices in Fig. 4.1 and Fig. 4.2 can be of crucial importance for the proper operation of the integrator in Fig. 4.3, especially when the maximum voltage supply is pushed down. The design equation derived in this section can be used to predict the deviation of the real tuning voltage V_{inr}, from the chosen value V_b as a function of the supply voltage as well as the common mode and the gain of the feedback amplifier.

The current flowing through transistors M1 and M2 is the same. Transistor M2 has to be biased in the strong inversion saturation region for the proper operation. Under these conditions, using equations (2.7) and (2.8)

$$\beta_1 (V_{\text{FG}} - V_{\text{Tn}}) V_b \approx \frac{\beta_2}{2} (V_{\text{DD}} - V_{\text{out(N2)}} - |V_{\text{Tp}}|)^2 \qquad (4.5)$$

where $V_{\text{out(N2)}}$ is the output of the tuning amplifier and as such it is a nonlinear function of $V_b - V_{\text{inr}}$[17]. However, as a first approximation the first two terms of a Taylor series expansion are good enough to represent this function since $V_{\text{inr}} \cong V_b$:

$$V_{\text{out(N2)}} = f(V_b - V_{\text{inr}}) \approx V_{\text{out(N2)}Q} - g_{\text{mf}} R_{\text{outf}} (V_b - V_{\text{inr}}) \qquad (4.6)$$

The subscript f refers to the feedback block N2. $g_{\text{mf}} R_{\text{outf}}$ represents its input/output gain[18]. $V_{\text{out(N2)}Q}$ is the operating point at the output when its differential input is zero,

---

[17] Neglecting common mode effects.

[18] The tuning amplifier N2 is just a differential pair. Its gain is the product of its transconductance and the output resistance.

i.e. $V_b = V_{inr}$. Combining eq. (4.5) and eq. (4.6), the expression for V_{inr} is

$$V_{inr} = V_b + \frac{1}{g_{mf}R_{outf}}(D - \sqrt{BV_b}) \tag{4.7}$$

where

$$D = V_{DD} - V_{out(N2)Q} - |V_{Tp}| \tag{4.8}$$

$$B = \frac{2.\beta_1}{\beta_2}(V_{FG} - V_{Tn}) \tag{4.9}$$

Equation (4.7) shows how the function which relates the tuning voltage, V_b, with the voltage, V_{inr}, fixed by the N2 block is not an exact equality, especially when V_b decreases, which might increase the complexity of the tuning mechanism for low transconductance values. This effect can be made less significant by increasing the gain of N2; however, doing this could cause instability as will be shown in the next chapter.

4.3.1.1   The threshold voltage shift

All the discussion and equations derived in previous sections are based on the assumption that all transistors are operating in the strong inversion region, which means that the condition $|V_{GS}| > |V_T|$ is satisfied (see eqs. (2.6–2.9)) for all the devices. If applied to transistor M2 in each N1 block, this means:

$$V_{out(N2)} < V_{DD} - |V_{Tp}| \tag{4.10}$$

As an example, in a technology with $|V_{Tp}| \approx 0.8$ V, and for a supply voltage of 1.5 V, eq. (4.10) would restrict the voltage at the output of the tuning amplifier to a maximum value of 0.7 V. But, besides, nodes V_R and $V_{out(N2)}$ in Fig. 4.2 should have close operating points for the sake of accuracy. Also, the n-channel MOS transistors M_n should work in the strong inversion region for matching and speed reasons, which means that their gate voltage has to be above their threshold voltage ($V_G > V_{Tn} \approx 0.8$ V). Both conditions cannot be met simultaneously if the mirror is implemented with normal n-channel MOS transistors. However, if FGMOS devices are used instead, as shown in Fig. 4.2, the design constraint for V_R is

$$V_R > \frac{C_T}{C_R}V_{Tn} - \frac{C_{VDD}}{C_R}V_{DD} \tag{4.11}$$

where C_R is the value of the capacitance connected to the node V_R and C_{VDD} is the capacitance of the other FGMOS input. This input is connected to the maximum voltage supply, V_{DD}, in order to shift the threshold voltage and thus relax the voltage requirements. Equation (4.11) neglects the effect of the parasitic capacitances on the FG[19] voltage. This approach will be followed from now on, unless the opposite is asserted. The reader can find out more about the design of the current mirror with FGMOS in Chapter 2.

---

[19] This is valid whenever they are much smaller than the input capacitances.

The threshold voltage has also been shifted in the input transistor of the N1 block by connecting one input capacitance in M1 to V_{DD}. This is done with the aim to improve the input range of the circuit for the chosen common mode at the input so that the condition of having the transistors working in the strong inversion region is met for a wider range of input voltages. A more exhaustive study on how to choose this capacitance can be found in Chapter 3, Section 3.3.5.

4.3.1.2   The cascode output stage. Output swing

The low frequency performance of the integrator in Fig. 4.3 can be improved by adding two-input FGMOS cascode transistors at the output stage (M_{c1}, M_{c1r}, M_{c3}, M_{c3r}). The reason for choosing FGMOS instead of normal MOS devices is twofold: on one hand, one input of each transistor can be used as part of the common mode feedback mechanism, needed because of the fully differential nature of the topology; on the other hand, the second input can be used to shift the effective threshold voltage hence making the low voltage constraint less restrictive. The common mode feedback strategy will be explained in the next section. Let us now explain the benefits of using the FGMOS instead of MOS devices in this particular topology.

The cascode transistors have to operate in the saturation region, preferably in strong inversion, in order to maximise the speed. The conditions for this are (according to eqs. (2.6–2.9))[20]

$$V_{FGcn} - V_{s1p} > V_{Tn} \tag{4.12}$$

$$V_{s3p} - V_{FGcp} > |V_{Tp}| \tag{4.13}$$

where V_{FGcn} and V_{FGcp} are the voltages at the FGs of the n-channel and p-channel cascode transistors, respectively. The previous equations can be rewritten as

$$V_{outcm} > V'_{Tn} + V_{s1p} \tag{4.14}$$

$$V_{outcm} < V_{s3p} - V'_{Tp} \tag{4.15}$$

where V_{outcm} is the voltage at the output of the common mode feedback circuit, V_{s1p} and V_{s3p} are the voltages at the sources of the n-channel and p-channel cascode transistors, respectively, and V'_{Tn}, V'_{Tp} are their effective threshold voltages which in this case are given by

$$V'_{Tn} = \frac{C_T}{C_{CMn}} \left[V_{Tn} - \frac{C_{VDDf}}{C_T} (V_{DD} - V_{s1p}) \right] \tag{4.16}$$

$$V'_{Tp} = \frac{C_T}{C_{CMp}} \left[|V_{Tp}| - \frac{C_{gndf}}{C_T} V_{s3p} \right] \tag{4.17}$$

---

[20] A similar expression can be obtained for the negative branch.

C_{VDDf}, C_{gndf} are the capacitances connected to V_{DD} and ground in the n-channel and p-channel cascode transistors, respectively, and C_{CMn}, C_{CMp} are the capacitances connected to V_{outcm}.

According to equations (4.14) and (4.15) V_{outcm} has to be high enough in order to keep the n-channel transistor in the strong inversion region, but at the same time low enough, so the p-channel device remains in that region as well. However, as the voltage supply decreases the limits for the bias voltage move closer to the supply voltages. Because of this the generation of a suitable bias signal becomes a more complex task. The FGMOS can simplify it by shifting the effective threshold voltages with adequate values of C_{VDDf} and C_{gndf} according to eqs. (4.16) and (4.17) These expressions show how the reduction of the effective threshold voltages is directly proportional to the value of the weights C_{VDDf}/C_T and C_{gndf}/C_T for the n-channel and p-channel devices, respectively. Nevertheless, the designer should bear in mind that there is a price to pay for this, since the input transconductances of the FGMOS transistors also decrease proportionally, and hence the output resistance. Still, as long as the effective input transconductance is kept larger than the output conductance for both FGMOS devices, it will be beneficial to use them at the output stage.

4.4   The Common Mode Feedback Block (CMFB)

The characteristic of fully differential structures is that signals are differential instead of being referred to ground. This property, however, is also the origin of one of their more important drawbacks: the common mode voltage of fully differential structures cannot be stabilised by the differential feedback. Thus, in order to stabilise the operating point an additional Common-Mode Feedback circuit (CMFB) has to be included in the topology [160].

The design of a CMFB can be difficult and it increases the complexity of the transconductor as the final topology has two signal paths: one defining the differential transfer function of the system and another one for the common mode signals. The circuit must detect the common signals as fast as possible in order to prevent them from affecting the differential ones. This should be achieved with a simple block, that ideally would not affect the overall integrator performance. Besides if the total power is a limiting factor, the circuit should consume as little power as possible.

Several CMFB implementations have been reported. Some of them use a capacitor–resistor network to detect the common mode level [148]. Although this technique is very efficient because of its zero power consumption and the high linearity, it requires a large area and it might also reduce drastically the output resistance of the transconductor. An alternative to this is to use transistors biased in the ohmic region [211]. The problem of this technique is that it reduces the transconductance and the bandwidth of the loop. The most common technique is the use of a differential pair in which the transistors at the output are saturated, and for low differential voltages, the AC-signal at the sources is equal to the common mode input signal. The problem with this structure is that it can be strongly nonlinear when the differential signal exceeds a certain value. It also degrades the differential impedance.

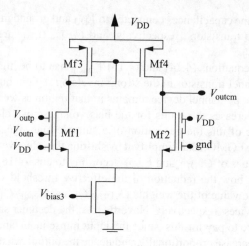

Figure 4.4   Common Mode Feedback Block (CMFB)

The circuit in Fig. 4.3 is a fully differential structure and as such it requires a CMFB in order to correct variations in the common mode at the output of the transconductor. An example of a block that could be used as CMFB in this particular design is shown in Fig 4.4. It consists of a differential pair with FGMOS type transistors at the input and a MOS current mirror as a load. The mean value of the single positive and negative outputs is sensed at the gate of one of the input FGMOS transistors and compared to a reference voltage generated at the gate of the other FGMOS device. If they are not equal the difference is amplified at V_{outcm} and fed back to both cascode transistors. This unbalances their currents in opposite directions, which forces the common mode at the output to move towards the reference value in order to equalise the sinked and sourced currents within the same branch.

The reference voltage in Mf2 (see Fig. 4.4) can be generated either by directly connecting the voltage supplies to two properly weighted inputs or with a biasing circuit. The first option would affect the PSRR as it will be shown later on.

Also, an extra input can be added to each of the input FGMOS devices with the purpose of reducing the effective threshold voltages in order to keep the input transistors in the strong inversion region (for matching and speed reasons) even in cases when the common mode at the output is below the nominal threshold voltage.

4.5 Small signal considerations

Small signal analysis of the cells forming the transconductor is presented in this section and several issues, important for the design of this circuit, are brought to the reader's attention.

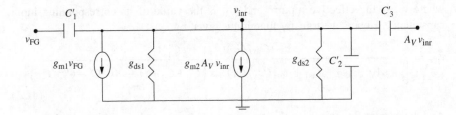

Figure 4.5   Small signal equivalent circuit for N1

Hence, for example, the effect of the gate to drain parasitic capacitance will be described and quantified. It will be shown how the latter introduces an error in the transconductance, causes a zero in the transfer function and affects the circuit stability.

Also, small signal analysis will prove how the zero of the N2 block is transmitted to the output current. The frequency at which this zero happens depends on the values of the input capacitances for the FGMOS transistors in N2.

Furthermore, the ratio between the gate to drain parasitic capacitances and the total capacitance in N2 FGMOS transistors influences N2 dominant pole and this, together with the value of the capacitance used to shift the effective threshold voltage (connected to V_{DD}), can again affect the transconductor stability.

Finally, the CMRR is going to suffer from the mismatch between the FGMOS transistors input capacitances. This effect will also be analysed for low frequencies.

4.5.1   The transconductance cell (N1)

A small signal equivalent circuit for Fig. 4.1 is shown in Fig. 4.5. A_V is the small-signal gain of the feedback block N2. The parasitic capacitances are given by

$$C_1' = C_{GD1} \tag{4.18}$$

$$C_2' = C_{DB1} + C_{DB2} + C_{LN2} \tag{4.19}$$

$$C_3' = C_{GD2} \tag{4.20}$$

C_{GDi}, C_{DBi} are the gate to drain and drain to bulk parasitic capacitances for transistor Mi. C_{LN2} represents the capacitive load due to the connection with the tuning amplifier N2. The small signal voltage at the FG is

$$V_{FG} = \frac{C_{in}}{C_T} v_{in} + \frac{C_1'}{C_T} v_{inr} \tag{4.21}$$

where v_{in} is the effective input[21] and C_{in} is the value of its corresponding input capacitance. Hence, the transfer function is given by

$$v_{inr} = \frac{-(C_{in}/C_T)(g_{m1} - sC'_1)v_{in}}{\left[g_{m2}A_V + g_{ds2} + g_{ds1} + g_{m1}(C'_1/C_T) + s\left(C'_1 + C'_2 + (1 - A_V)C'_3 - C'^2_1/C_T\right)\right]}$$

(4.22)

And the current flowing through a transistor M3 connected, as a mirror, to M2:

$$i_{out} = \frac{(C_{in}/C_T)g_{m3}A_V(g_{m1} - sC'_1)v_{in}}{\left[g_{m2}A_V + g_{ds2} + g_{ds1} + g_{m1}(C'_1/C_T) + s\left(C'_1 + C'_2 + (1 - A_V)C'_3 - C'^2_1/C_T\right)\right]}$$

(4.23)

Modelling the frequency response of the feedback circuit (A_V) as a single pole – single zero function, eq. (4.23) can be rewritten as

$$i_{out} = \frac{(C_{in}/C_T)g_{m3}A_0(1 - s/z)(g_{m1} - sC'_1)v_{in}}{\left[(g_{m2} - sC'_3)A_0(1 + s/z) + \left[g_{ds2} + g_{ds1} + g_{m1}(C'_1/C_T) + sC_{eq}\right](1 + s/p)\right]}$$

(4.24)

where A_0, p and z are the DC gain, pole and zero of the tuning block (N2), respectively. Also:

$$C_{eq} = C'_1 + C'_2 + C'_3 - \frac{C'^2_1}{C_T}$$

(4.25)

The feedback circuit has a couple of important effects on the frequency response of the transconductor core block. On one hand, it affects the value of the DC gain which now is given by

$$i_{outr} = \frac{i_{outi}}{1 + \varepsilon}$$

(4.26)

where the subscripts r and i refer to the real and ideal DC responses and ε quantifies the DC error:

$$\varepsilon = \frac{g_{m1}}{g_{m2}} \frac{C'_1}{A_0 C_T} + \frac{(g_{ds2} + g_{ds1})}{A_0 g_{m2}}$$

(4.27)

Equations (4.26) and (4.27) reveal how the larger the DC gain of the feedback circuit is the smaller is the error. Also, the first term in eq. (4.27) is proportional to the ratio between the gate to drain parasitic (C'_1) and the total capacitance, which means that if this ratio is smaller the error will be smaller too.

On the other hand, the numerator of eq. (4.24) shows how the zero of the feedback amplifier is transmitted to the output current, so this amplifier should be designed in such a way that its zero remains far away from the operating frequency band of the circuit. Apart from this zero, there is another one caused by the coupling between M1

---

[21] The input for which the transfer function is calculated.

drain and the FG. This means that in order to achieve the desired operating frequency range a trade-off between the maximum transconductance, the power and the area has to be met. On one hand, increasing the transconductance will increase the power as well as the error in eq. (4.27), but it will also improve the speed. On the other hand, the error can be reduced by increasing the total capacitance seen by the input FG (without changing the C_{in}/C_T ratio), but this would also increase the circuit area.

Another conclusion can be extracted from the previous results analysing the second order polynomial in the denominator of eq. (4.24): the circuit is only conditionally stable, this is whenever any of the conditions shown in eq. (4.28) or eq. (4.29) happen, the circuit will oscillate. Hence, the tuning amplifier must be designed in such a way that none of them ever occurs:

$$\left[\frac{g_{m2}A_0}{z} + \left(C'_1 + C'_2 + (1 - A_0)C'_3 - \frac{C'^2_1}{C_T} \right) + g_{m1}\frac{C'_1}{C_{Tp}} + \frac{(g_{ds1} + g_{ds2})}{p} \right] < 0 \tag{4.28}$$

$$\left(\frac{\left(C'_1 + C'_2 + C'_3 - C'^2_1/C_T \right)}{p} - \frac{C'_3 A_0}{z} \right) < 0 \tag{4.29}$$

The analysis of the tuning amplifier frequency response is carried out in the following section.

4.5.2 The tuning amplifier (N2)

The small signal equivalent circuit for the tuning amplifier (N2) is shown in Fig. 4.6[22], where

$$C_{02} = C_{VDD} + 2C_{GSn} + 2C_{GBn} \tag{4.30}$$

$$C_{22} = C_{DBn} + C_{DBp} + C_{GDp} + C_L \tag{4.31}$$

$$C_{42} = C_{DBn} + C_{DBp} \tag{4.32}$$

Subscripts n and p refer to the name of the device. g_{dsbias} accounts for the real conductance of the current source that generates I_{B2}. C_L refers to a capacitive load at the output:

$$\frac{v_{out(N2)}}{v_{inr}} = \frac{(g_{mp}/2)(s/p_1 + 1)}{A_1 s^2 + B_1 s + D_1} \tag{4.33}$$

---

[22] The pMOS transistors are assumed to be in separate wells. Also, for the sake of simplicity, only the most important capacitances have been considered. Therefore, the following analysis will only be valid for mid-frequencies, but this is good enough to obtain the most important conclusions.

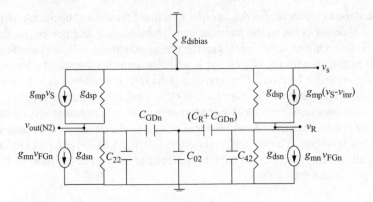

Figure 4.6   Small signal equivalent circuit for N2

$$A_1 \approx \frac{1}{p_1}\left(C_{22} + C_{GDn} - \frac{C_{GDn}^2}{C_T}\right) - C_{GDn}\frac{(C_R + C_{GDn})}{C_T}\frac{a_0}{z_1} \tag{4.34}$$

$$B_1 \approx \frac{(g_{mn}(C_{GDn}/C_T) + g_{dsn} + g_{dsp}/2)}{p_1} + \frac{g_{mn}(C_R + C_{GDn})}{C_T}\frac{a_0}{z_1}$$
$$+ \left(C_{22} + C_{GDn} - \frac{C_{GDn}^2}{C_T}\right)a_0\frac{C_{GDn}}{C_T}(C_R + C_{GDn}) - \frac{g_{dsp}}{2}\frac{a_0}{z_1} \tag{4.35}$$

$$D_1 \approx g_{mn}a_0\frac{(C_R + C_{GDn})}{C_T} + g_{mn}\frac{C_{GDn}}{C_T} + g_{dsn} + \frac{g_{dsp}}{2} - \frac{g_{dsp}}{2}a_0 \tag{4.36}$$

where in this case, the total capacitance seen by the FG is

$$C_T \approx 2C_{GDn} + C_{02} + C_R \tag{4.37}$$

a_0, p_1 and z_1, are the gain, transmission pole and zero from $v_{out(N2)}$ to v_R, respectively. The values of these parameters are collected in equations (4.38), (4.39) and (4.40):

$$a_0 = \frac{g_{dsn} + 2g_{mn}(C_{GDn}/C_T)}{g_{dsn} + 2g_{mn}((C_{GDn} + C_R)/C_T)} \tag{4.38}$$

$$z_1 = \frac{g_{dsn} + 2g_{mn}(C_{GDn}/C_T)}{(C_{22} + C_{GDn}) - C_{GDn}(C_R + 2C_{GDn})/C_T} \tag{4.39}$$

$$p_1 = \frac{g_{dsn} + 2g_{mn}((C_R + C_{GDn})/C_T)}{(C_{42} + C_R + C_{GDn}) - (C_R + C_{GDn})(C_R + 2C_{GDn})/C_T} \tag{4.40}$$

Therefore, N2 DC gain will be given by

$$A_0 \approx \frac{g_{mp}/2}{g_{mn}a_0((C_R+C_{GDn})/C_T)+g_{mn}(C_{GDn}/C_T)+g_{dsn}+g_{dsp}/2-(g_{dsp}/2)a_0}$$

(4.41)

As expected the increased output conductance of the load FGMOS devices affects the DC gain, A_0, which is now reduced if compared to the gain of the same topology built with normal MOS devices instead.

N2 dominant pole can be calculated using the following assumptions:

$$2g_{mn}\frac{(C_{GDn}+C_R)}{C_T} \gg g_{dsp}+g_{dsn}$$

(4.42)

$$C_{22} \gg \left(1-\frac{C_R}{C_T}\right)C_{GDn}$$

(4.43)

$$C_R \gg C_{GDn}$$

(4.44)

which gives the approximate value

$$p \approx \frac{2[g_{mn}a_0((C_R+C_{GDn})/C_T)+g_{mn}(C_{GDn}/C_T)+g_{dsn}+g_{dsp}/2-g_{dsp}a_0/2]}{3C_{22}}$$

(4.45)

The value of the zero is given by

$$z \approx -\frac{2g_{mn}}{C_T-C_R}$$

(4.46)

Hence, the oscillation conditions in eq. (4.29) will never happen, because of the negative value of z. Also eq. (4.28) can be rewritten as

$$\left(C_1'+C_2'+C_3'-\frac{C_1'^2}{C_T}\right)+g_{m1}\frac{C_1'}{C_{Tp}}+\frac{(g_{ds1}+g_{ds2})}{p} < \left[C_3'+\frac{g_{m2}}{2g_{mn}}(C_T-C_R)\right]A_0$$

(4.47)

The right-hand side in eq. (4.47) shows how in order to minimise the risk of oscillation the term relating the transconductor transconductance, the capacitance connected to V_{DD} in the current mirrors and N2 transconductance should remain as small as possible and the relationship between the values of the parasitic and load capacitances should be such that eq. (4.47) never occurs.

4.5.3   Common mode DC response

The integrator in Fig. 4.3 is a fully differential structure. Therefore, ideally, its common mode gain should be as small as possible. By replicating the small signal equivalent circuit shown in Fig. 4.5 for both differential branches, and performing a small signal analysis, the equation for the gain referred to one of the output branches (p) is[23]

$$\frac{v_{outp}}{v_{inCM}}\bigg|_{v_{ind}=0}$$
$$\approx \frac{[(C_{in}/C_T)_p(g_{mf}R_{outf})_pg_{m1}g_{ma3} - (C_{in}/C_T)_n(g_{mf}R_{outf})_ng_{m1r}g_{m3r}(g_{mal}/g_{m4r})]R_{outp}}{[g_{m2}(g_{mf}R_{outf})_p + g_{ds2} + g_{ds1} + g_{m1}(C_1'/C_T)_p][1 + A_{CMp}]}$$

(4.48)

R_{outp} is the name used for the output resistance in the output branch (p). Assuming that the p-channel FGMOS transistors are in separate wells, the value of R_{outp} is given by

$$R_{outp} = G_{outp}^{-1} \approx \Bigg[\frac{g_{dsFc3}g_{dsa3}}{[(C_{CMp}/C_T + C_{gndf}/C_T)_pg_{mc3} + g_{dsFc3}]}$$
$$+ \frac{g_{dsFc1}g_{dsa1}}{[(C_{CMn}/C_T + C_{VDDf}/C_T)_pg_{mc1} + g_{mbFc1} + g_{dsFc1}]}\Bigg]^{-1}$$
$$= [g_{dsFc3} \cdot R_{1p} + g_{dsFc1} \cdot R_{2p}]^{-1}$$

(4.49)

A_{CMp} depends on the CMFB. Its value is given by[24]

$$A_{CMp} \approx \frac{(C_{CMp}/C_T)_p(g_{mc3} \cdot R_{1p}) + (C_{CMn}/C_T)_p(g_{mc1}R_{2p})}{G_{outp}}\left[\frac{2g_{mf1}(C_{in}/C_T)}{g_{dsf4} + g_{dsFf2}}\right]_p$$
$$= \frac{A_{CM1p}}{G_{outp}}\left[\frac{2g_{mf1}(C_{in}/C_T)}{g_{dsf4} + g_{dsFf2}}\right]_p$$

(4.50)

The differential gain can be derived by assuming that both branches are identical and $R_{outp} = R_{outn} = R_{out} = G_{out}^{-1}$ [25]:

$$\frac{v_{out}}{v_{ind}}\bigg|_{v_{inCM}=0} \approx G_m R_{out}$$

(4.51)

The Common Mode Rejection Ratio (CMRR) is given by the ratio between the differential gain in eq. (4.51) and the difference between the common mode gains of both output branches assuming mismatch (eq. (4.48)):

$$CMRR = \frac{[v_{out}/v_{ind}]\,|_{v_{inCM}=0}}{[((v_{outp}/v_{inCM})\,|_{v_{ind}=0} - (v_{outn}/v_{inCM})\,|_{v_{ind}=0})]}$$

(4.52)

---

[23] The subscript p in $g_{mf}R_{outf}$ is used to make reference to the N2 amplifier corresponding to the p branch. Also, in general when either the subscript p or n is added to the name of a parameter it will refer to the p or n branch, respectively.

[24] The value of the input capacitance C_{in} has been assumed to be the same as in the CMFB.

[25] For the sake of clarity. As its value is supposed to be very high this will not affect much the final result.

The main conclusion drawn from this analysis is that the use of FGMOS transistors can have a detrimental effect on the CMRR. This is because the common mode gain at the output increases as the gain of the CMFB decreases. Having FGMOS transistors as cascode devices will reduce $A_{CM(n,p)}$ due to the reduction of the transconductance and the increase of the total output conductance through g_{dsFci}. However, if the FGMOS parasitic capacitances in the CMFB are small enough (compared with the total capacitance), the gain of the block can still be large enough to meet the specifications and compensate for the compression of the signals at the FGs. It is also worth pointing out that the mismatch between the input capacitances in the FGMOS transistors is going to affect the value of the common mode output gain. As an example, for a 1 per cent mismatch in the input capacitances, $A_{CM(n,p)} = 10$, $k = 1$, and values of conductances much smaller than the values of the transconductances, the magnitude of the CMRR will be higher than 60 dB.

4.6   Second order effects

Two typical analog filter design specifications are linearity and noise. Linearity limits the value of the largest useful signal that can be handled by the filter while noise establishes the limit for the smallest. DR is the ratio between these two magnitudes, and it is a crucial figure in LV/LP analog design. The maximum input signal that can be applied to a circuit depends directly on the Total Harmonic Distortion requirement (THD) for the system [212]. This section presents a study of both the THD and the noise in the transconductor block together with a discussion of the repercussions of using the FGMOS transistors in the values of these parameters.

4.6.1   Total harmonic distortion

When a sinusoidal waveform is applied to a linear time-invariant system, the output will also be a sinusoidal waveform at the same frequency, although, possibly with different magnitude and phase values. If this input is applied to a nonlinear system the output will have extra frequency components (harmonics) at frequency values which are multiples of the input frequency. The THD of a signal is defined as the ratio of the total power of the second and higher order harmonic components to the power of the fundamental harmonic:

$$\text{THD} = \frac{\sqrt{(H_{D2}^2 + H_{D3}^2 + H_{D4}^2 + \cdots)}}{H_{D1}} \times 100 \tag{4.53}$$

H_{D1}, H_{Di} are the amplitudes of the fundamental and ith-harmonic components, respectively.

A problem related to the use of THD as a filter characterisation parameter is that often harmonic components fall in the stopband of the filter, and thus the THD value can appear as falsely improved. In fact, the filter linearity is worse when higher input signal frequencies are applied due to nonlinear capacitances or nonlinear signal cancellation. A better way of measuring the linearity near the upper passband edge is by using the intermodulation test (IM3). This test is actually the only one that works for bandpass filters [212]. This section analyses the main sources of non-linearity in the integrator circuit. Also, an approximate equation for both, the THD and IM3 as a function of different design parameters is derived.

Equation (4.2) was obtained assuming that the drain current of the input FGMOS transistor in N1 was a perfect linear function of V_{DS}. Also, β_1 was supposed to be a signal-independent parameter. However, none of these assumptions is really true, and if non-ideal effects are taken into account the output of the transconductor block will not be an exact linear function of the voltage.

An equation that fits better the real drain current in the transistor is

$$I_D \approx \beta_1 \left[(V_{FG} - V_{Tn})V_D - \frac{V_D^2}{2} \right]$$
(4.54)

Also the mobility μ and consequently β_1 depend non-linearly on both V_{FG} and V_D, the dependence with V_{FG} being dominant. A function that models this approximately is [117][26]

$$\mu = \frac{\mu_0}{1 + \theta(V_{GS} - V_{Tn})}$$
(4.55)

where μ_0 represents the zero-field mobility of carriers and θ accounts for the effect of the vertical electric field.

An expression for the drain voltage can be obtained by considering that the currents flowing through transistors M1 and M2 in Fig. 4.1 are the same[27]:

$$V_D = \frac{-B' + \sqrt{B'^2 + 2(\beta_2/\beta_1)k_4^2[C_{GD1}/C_T - 1/2 - (\beta_2/2\beta_1)g_{mf}^2 R_{outf}^2]}}{2[C_{GD1}/C_T - 1/2 - (\beta_2/2\beta_1)g_{mf}^2 R_{outf}^2]}$$
(4.56)

where:

$$B' = \frac{C_{VDD}}{C_T}V_{DD} + \frac{C_{in}}{C_T}V_{inCM} - V_{Tn} + \frac{C_{in}}{C_T}\frac{V_{ind}}{2}$$
$$+ \frac{\beta_2}{\beta_1}g_{mf}R_{outf}(V_{DD} - V_{out(N2)Q} + g_{mf}R_{outf}V_b - |V_{Tp}|)$$
(4.57)

$$k_4 = V_{DD} - V_{out(N2)Q} + g_{mf}R_{outf}V_b - |V_{Tp}|$$
(4.58)

Assuming that the differential input signal is sinusoidal with an amplitude A for each branch, putting these expressions into the equations describing the currents

---

[26] V_{GS} is, as usual, the gate to source voltage. In this case the gate is floating.
[27] V_D stands for the drain voltage that was called V_{inr} in Fig. 4.1. V_{inCM} is the common mode at the input.

flowing through M1 and M2 and simplifying them using Taylor series expansions, the nonlinear eq. (4.59) results:

$$I_{\text{out}} \approx (c_0 d_1 + c_1 d_0)A + (c_0 d_3 + c_1 d_2 + c_2 d_1)A^3 \tag{4.59}$$

Where coefficients are defined as follows:

$$c_0 \approx \beta_{1i} \left[1 - \theta \left(\frac{C_{\text{VDD}}}{C_T} V_{\text{DD}} + \frac{C_{\text{in}}}{C_T} V_{\text{inCM}} - V_{\text{Tn}} \right) \right] \tag{4.60}$$

$$c_1 \approx -\beta_{1i} \theta \frac{C_{\text{in}}}{C_T} \tag{4.61}$$

$$c_2 \approx \beta_{1i} \theta^2 \left(\frac{C_{\text{in}}}{C_T} \right)^2 \tag{4.62}$$

β_{1i} represents the ideal β_1 and

$$d_0 = V_{\text{D}(V_{\text{ind}}=0)} \left(\frac{C_{\text{VDD}}}{C_T} V_{\text{DD}} + \frac{C_{\text{in}}}{C_T} V_{\text{inCM}} - V_{\text{Tn}} \right) - \left(\frac{1}{2} - \frac{C_{\text{GD1}}}{C_T} \right) V_{\text{D}(V_{\text{ind}}=0)}^2 \tag{4.63}$$

$$d_1 = \frac{C_{\text{in}} V_{\text{D}(V_{\text{ind}}=0)}}{C_T} - \left[\left(\frac{C_{\text{VDD}}}{C_T} V_{\text{DD}} + \frac{C_{\text{in}}}{C_T} V_{\text{inCM}} - V_{\text{Tn}} \right) \frac{C_{\text{in}}}{C_T} k_5 \right.$$
$$\left. -2 \left(\frac{1}{2} - \frac{C_{\text{GD1}}}{C_T} \right) \left(\frac{C_{\text{in}}}{C_T} k_5 V_{\text{D}(V_{\text{ind}}=0)} \right) \right]$$
$$\times \left[2 \left[\frac{C_{\text{GD1}}}{C_T} - \frac{1}{2} - \frac{\beta_2}{2\beta_1} g_{\text{mf}}^2 R_{\text{outf}}^2 \right] \right]^{-1} \tag{4.64}$$

$$k_5 = \left(1 - \frac{((C_{\text{VDD}}/C_T)V_{\text{DD}} + (C_{\text{in}}/C_T)V_{\text{inCM}} - V_{\text{Tn}} + (\beta_2/\beta_1)g_{\text{mf}}R_{\text{outf}}k_4)}{\sqrt{B'^2_{(V_{\text{ind}}=0)} + (2\beta_2/\beta_1)k_4^2 \left[C_{\text{GD1}}/C_T - 1/2 - (\beta_2/2\beta_1)g_{\text{mf}}^2 R_{\text{outf}}^2 \right]}} \right)$$
$$= 1 - \frac{B'_{(V_{\text{ind}}=0)}}{\sqrt{k_6}} \tag{4.65}$$

$$d_2 = \left[B_4' \times \left[\left(\frac{C_{\text{VDD}}}{C_T} \right) V_{\text{DD}} + \frac{C_{\text{in}}}{C_T} V_{\text{inCM}} - V_{\text{Tn}} \right.\right.$$
$$\left.\left. -2 \left(\frac{1}{2} - \frac{C_{\text{GD1}}}{C_T} \right) V_{\text{D}(V_{\text{ind}}=0)} \right] \left(\frac{C_{\text{in}}}{C_T} \right)^2 \left[8k_7 k_6^{3/2} \right]^{-1} \right.$$
$$- \frac{(C_{\text{in}}/C_T) k_5}{2k_7} \left[\frac{C_{\text{in}}}{C_T} + \frac{(1/2 - C_{\text{GD1}}/C_T)(C_{\text{in}}/C_T) k_5}{2k_7} \right] \tag{4.66}$$

$$k_7 = \left[\frac{C_{GD1}}{C_T} - \frac{1}{2} - \frac{\beta_2}{2\beta_1} g_{mf}^2 R_{outf}^2 \right] \tag{4.67}$$

$$B_4' = 2 \cdot k_4^2 \cdot \frac{\beta_2}{\beta_1} \cdot k_7 \tag{4.68}$$

$$d_3 = \frac{B_4' \, (C_{in}/C_T)^2}{8 k_7 k_6^{3/2}} \left[\frac{C_{in}}{C_T} + \frac{(1/2 - C_{GD1}/C_T)(C_{in}/C_T) k_5}{k_7} \right]$$

$$- \frac{B_4' \, (C_{in}/C_T)^3 \times B_{(V_{ind}=0)}'}{4 k_7 k_6^{5/2}} \times \left[\left(\frac{C_{VDD}}{C_T} V_{DD} + \frac{C_{in}}{C_T} V_{inCM} - V_{Tn} \right) \right.$$

$$\left. - 2 \times \left(\frac{1}{2} - \frac{C_{GD1}}{C_T} \right) \times V_{D(V_{ind}=0)} \right] \tag{4.69}$$

Since the structure is fully differential, it can be assumed that the THD is dominated by the third-order harmonic. If this is the case the THD is given by

$$\text{THD} \approx \frac{1}{4} \frac{(c_0 d_3 + c_1 d_2 + c_2 d_1)}{(c_0 d_1 + c_1 d_0)} A^2 \tag{4.70}$$

And, the third-order intermodulation value is

$$IM_3 \approx \frac{3}{4} \left(\frac{(c_0 d_3 + c_1 d_2 + c_2 d_1)}{(c_0 d_1 + c_1 d_0)} A^2 \right) \tag{4.71}$$

The following figures (Fig. 4.7 to Fig. 4.9) represent the THD as a function of various parameters, according to eq. (4.70). Figure 4.7 shows the THD versus two parameters – input amplitude and input weight[28]. As expected (see Chapter 2, Section 2.3.3), the THD decreases as the signal is further attenuated at the FG. However, even for a relatively high value of the input weight $(C_{in}/C_T) = 0.5$, the value of the THD is still lower than 0.6 per cent.

Also, Fig. 4.8 shows how the increase of the drain to gate capacitive coupling contributes slightly to the reduction of the THD as C_{in}/C_T increases. The reason for this is that the nonlinear term in the current expression is now multiplied by $(0.5 - C_{GD}/C_T)$, instead of 0.5, which reduces the partial derivative of the output current with respect to V_{DS}^2.

Figure 4.9 shows how the THD increases with the input signal amplitude for very low values of the tuning voltage[29]. This is due to the finite gain of the tuning amplifier N2. The same happens for large values of V_b. This is because the input transistor approaches the saturation region and the behaviour of the current becomes more nonlinear. Hence, the maximum tuning voltage will be limited by the operating region of the input transistor while the minimum by the tuning amplifier.

---

[28] The values of the parameters used to obtain these surfaces are the same as for the design example described at the end of the chapter.

[29] Always under the assumption that the input transistor is operating in the ohmic region.

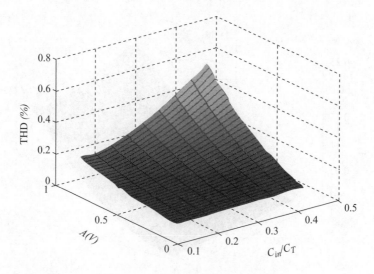

Figure 4.7   THD as a function of an input amplitude and C_{in}/C_T

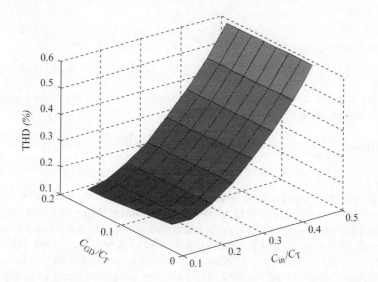

Figure 4.8   THD versus C_{GD}/C_T and C_{in}/C_T

The previous analysis was carried out under the assumption that the fully differential structure is completely ideal, which, in general, is not true due to the mismatch between both differential branches. Mismatch will increase the THD since the second-order harmonic will not be zero anymore. The effect can be quantified using the same derivations as before but considering different coefficients for each branch. This will

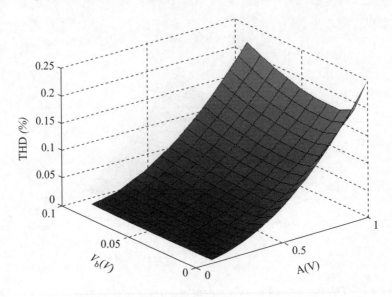

Figure 4.9   THD versus the amplitude of the input signal and the tuning voltage

add the following second-order term to the expression of the current[30]:

$$H_{D2} \approx [(c_{0p}d_{2p} + c_{1p}d_{1p} + c_{2p}d_{0p}) - (c_{0n}d_{2n} + c_{1n}d_{1n} + c_{2n}d_{0n})]A^2 \quad (4.72)$$

The second-order distortion term defined as the ratio between the second- and first-order harmonic will then be

$$HD2 = \frac{H_{D2}}{H_{D1}} \approx \frac{[(c_{0p}d_{2p} + c_{1p}d_{1p} + c_{2p}d_{0p}) - (c_{0n}d_{2n} + c_{1n}d_{1n} + c_{2n}d_{0n})]}{2(c_{0n}d_{1n} + c_{1n}d_{0n} + c_{0p}d_{1p} + c_{1p}d_{0p})}A$$

$$(4.73)$$

HD2 is represented in Fig. 4.10 for an input capacitance weight of 0.25 and a drain to gate parasitic capacitance ratio of 0.05, as a function of the input signal amplitude and the percentage of mismatch in the input capacitances of the n and p branches. Although mismatch in other parameters could also have an effect on the THD, these are the most significant ones. Table 4.1 shows the value of parameters used in the calculations. Experimental results of THD are presented from Fig. 4.18 to Fig. 4.22.

4.6.1.1   Noise

The lower limit of the signal that can be handled by the circuit is determined by the noise. The noise of the transconductor in Fig. 4.3 has three main contributors: the main cell, the tuning amplifier and the common mode feedback circuit.

---

[30]   Subscripts p and n refer to the positive and negative branches.

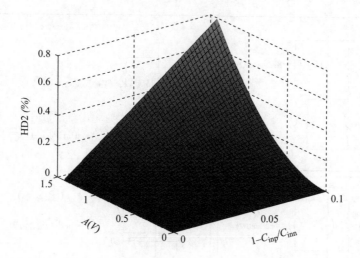

Figure 4.10 Second-order harmonic versus the amplitude of the input signal and the percentage of mismatch

Table 4.1 Design parameters

θ	0.234
$g_{mf}R_{outf}$	76
C_{GD}	8 fF
V_{inCM}	0.75 V

If both single branches in the circuit are assumed to be identical, the noise can be obtained by analysing the small signal equivalent circuit for one of them and multiplying the result by a factor of two. As a first approximation, the small signal circuit with all noise sources included is shown in Fig. 4.11. Although the subscript r in certain transconductance parameters makes reference to the other single branch, the effect of the mismatch will be ignored, which is equivalent to assuming that $g_{mi} \approx g_{mir}$. The schematic for the CMFB has not been drawn for the sake of clarity, as it is also similar to the N2 block. In the N2 block $|S'_{p1}| = |S'_{p2}| = S'_p$ and $|S'_{n1}| = |S'_{n2}| = S'_n$ and in general for transistor Mi

$$S_i = S'^2_i = 4KTg_{mi}\left[\frac{\rho g_{mi}}{W_i L_i f} + \gamma\right] \tag{4.74}$$

where ρ and γ depend on the technology and are different for n- and p-channel devices [117,213]. An approximate expression of the equivalent noise at the input valid for

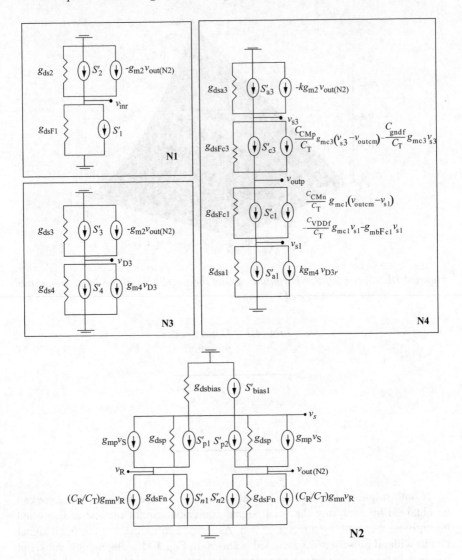

Figure 4.11   *Small signal equivalent circuit for noise calculations*

low frequencies is given by

$$\frac{\overline{v_{\text{ind}}^2}}{\Delta f} \approx 2 \left[\frac{(S_{a1} + S_{a3})}{G_m^2} + \frac{k^2(S_3 + S_4)}{G_m^2} + \frac{S_{c3}(R_1)^2 + S_{c1}(R_2)^2}{G_m^2} \right]$$

$$+ \frac{2k^2 g_{m2}^2}{G_m^2} \left[\frac{(g_{mp}^2/(g_{dsFn} + g_{dsp})^2)(S_1 + S_2)}{(g_{dsF1} + g_{ds2} + g_{m2}(g_{mp}/(g_{dsFn} + g_{dsp})))^2} \right] \qquad (4.75)$$

The noise generated by the CMFB is a common mode signal that ideally gets rejected if both positive and negative branches are identical. Because of this its value has been neglected from eq. (4.75).

The noise transfer functions for the tuning amplifier and the CMFB are given by

$$\frac{\overline{v^2_{out(N2)}}}{\Delta f} \approx \frac{2(S_p + S_n)}{(g_{dsp} + g_{dsFn})^2} \tag{4.76}$$

$$\frac{\overline{v^2_{outcm}}}{\Delta f} \approx \frac{2(S_{f1} + S_{f3})}{(g_{dsf3} + g_{dsFf1})^2} \tag{4.77}$$

The conclusions that can be extracted from the analysis are the following:

1. The FGMOS in the tuning amplifier affects its noise performance in the following way: the noise power spectral density is reduced as a consequence of the capacitive coupling between the drain and gate in the load current mirrors. Nevertheless, this is just an added term in the denominator of eq. (4.76), so ultimately its effect will depend on how small the output conductances are in comparison to it.
2. The same is applicable to the CMFB.
3. Regarding the total equivalent noise at the input of the transconductor, eq. (4.75) shows that the main noise contributors are those transistors that realise the transconductance function together with their mirrored counterparts (M1, M2, M3, M4, Ma1 and Ma3, and their symmetrical devices in the other branch). A strategy to deal with this could be to increase their area. However, this has to be done with precaution for the input FGMOS devices, since their parasitic capacitances would increase as well which might have other negative consequences (as it is being described along the book) and affect the operating range. As expected the noise is proportional to $(C_T/C_{in})^2$.

4.7 Power Supply Rejection Ratio (PSRR)

As previously shown in a number of sections of the book, the supply voltages can directly be used to shift the voltage at the gate of FGMOS devices, or what is the same, to change the effective threshold voltages. However, this might have a negative effect on the Power Supply Rejection Ratio (PSRR) of the circuit which is next analysed.

Figure 4.12 shows a small signal equivalent circuit used to calculate the PSRR. In order to simplify the notation for this circuit:

$$g'_{mi} = (C_{VDD}/C_T)g_{mi}, \quad g^{CM}_{mc1} = (C_{CMn}/C_T)g_{mc1},$$

$$g'_{mc1} = (C_{VDDf}/C_T)g_{mc1}, \quad g'_{mc3} = (C_{gndf}/C_T)g_{mi},$$

$$g^{CM}_{mc3} = (C_{CMp}/C_T)g_{mc3}, \quad g_{mgi} = \left(C_{VDD}g_{mi} + \sum_{i=1}^{N} C_i g_{mi}\right) \bigg/ C_T \tag{4.78}$$

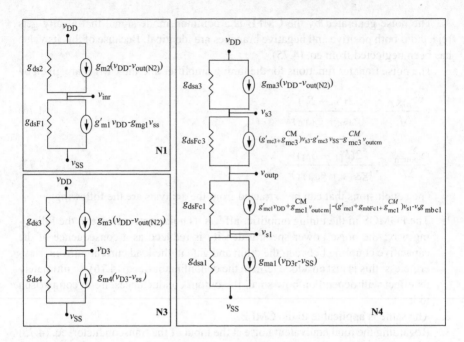

Figure 4.12   Small signal equivalent circuit for PSRR calculations

and v_{DD}, v_{SS} refer to the variations of the positive and negative supply voltage, respectively. Assuming that the single branches are not identical due to the mismatch, the following expression for the PSRR+[31] results:

$$\text{PSRR+} \approx \frac{G_m R_{out}}{(v_{out}/v_{DD}|_{V_{ind}=0})} \tag{4.79}$$

$$\left.\frac{v_{out}}{v_{DD}}\right|_{v_{ind}=0} \approx -R_{outp}g_{ma1}\frac{g_{m3r}}{g_{m4r}}(1-a'_{0n}) + R_{outn}g_{ma1r}\frac{g_{m3}}{g_{m4}}(1-a'_{0p})$$

$$- (A_{CM2p}R_{outp} - A_{CM2n}R_{outn}) - R_{outp}g_{ma3}a'_{0p} + R_{outn}g_{ma3r}a'_{0n}$$

$$- (A_{CM1p}R_{outp} - A_{CM1n}R_{outn})\left.\frac{v_{outcm}}{v_{DD}}\right|_{v_{out\,CM}=0}$$

$$+ g_{ma3}R_{outp} - g_{ma3r}R_{outn} \tag{4.80}$$

[31] R_{outp} and R_{outn} refer to the output resistance of the positive and negative branch, respectively. Their expression can be obtained from eq. (4.49) by interchanging the subscripts.

A_{CM1p} was previously defined in eq. (4.50), and A_{CM2p} is given by

$$A_{CM2p} = \left(\frac{C_{VDDf}}{C_T} \right)_p g_{mc1} R_{2p} \tag{4.81}$$

a'_{0p} depends on the tuning amplifier as

$$a'_{0p} = \frac{(g_{m2} - g_{m1} + g_{ds2})(g_{mf} R_{outf})_p}{[g_{dsF1} + g_{ds2} + g_{m2}(g_{mf} R_{outf})_p]} + \left. \frac{v_{out(N2)}}{v_{DD}} \right|_{v_{inr}=0} \tag{4.82}$$

A_{CM1n}, A_{CM2n} and a'_{0n} can be obtained from equation (4.50) and equation (4.81) just by replacing the subscripts by those corresponding to the other branch. The second term is related to V_{DD} rejection ratio as

$$\left. \frac{v_{out(N2)}}{v_{DD}} \right|_{v_{inr}=0} = \frac{C_{VDD}}{C_T} \frac{g_{mn}}{g_{dsFn} + g_{dsp}} = - \left(\frac{g_{mn}}{g_{dsFn} + g_{dsp}} \times \frac{1}{PSRR+_{(N2)}} \right) \tag{4.83}$$

Equation (4.83) shows how N2 PSRR+ is seriously degraded if V_{DD} is used to shift the threshold voltage of the load current mirror with the FGs connected together. A way to compensate for this would be not to connect the FGs, which would dramatically improve this value. Nevertheless, this would affect the accuracy and the output resistance of the mirror. Another solution, if the PSRR+ is an issue, would be to use a separate voltage source, although, the latter might increase the power consumption.

A small signal equivalent circuit for PSRR calculations in the CMFB is shown in Fig. 4.13. In this case, the output gain with respect to the positive voltage supply is[32]

$$\left. \frac{v_{outcm}}{v_{DD}} \right|_{v_{outCM}=0}$$

$$\approx \frac{C_{in}}{C_T} \frac{g_{mf1}}{g_{dsFf1} + g_{dsf4}} \times \frac{1}{PSRR+_{CMFB}}$$

$$= (g_{dsFf2} + g_{dsf4})^{-1} \left[[g_{mf3} - g_{mf4}] + \left(1 + \frac{g_{mf4}}{g_{mf3}} \right) \right.$$

$$\left. \times \frac{(g_{mf2} + g_{mbf2})g_{mf1}(C_{VDD}/C_T)_{f1} - (g_{mf1} + g_{mbf1})g_{mf2}(C_{VDD}/C_T)_{f2}}{g_{mf2} + g_{mbf2} + g_{mf1} + g_{mbf1}} \right] \tag{4.84}$$

where again the mismatch has been neglected in the expression of the differential gain. This is justified for two reasons: first, it increases the clarity, and second, because the differential gain is expected to be high and thus this assumption is not going to affect much the final result. C_{VDD}/C_T is the weight associated with V_{DD} in the input

---

[32] Assuming $g_{dsFf} \ll (C_{VDD}/C_T)g_{mf}$.

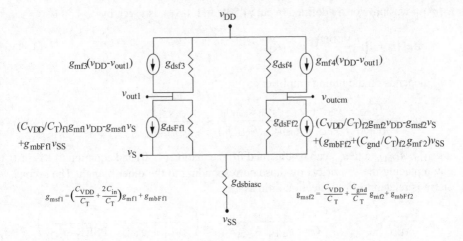

Figure 4.13　Small signal equivalent circuit for PSRR calculations in the CMFB

transistors. The weights are expected to be different because of the mismatch, and that is why subscripts f_1 and f_2 have been added in their names.

In order to have an idea of the order of magnitude predicted by eq. (4.84), and also isolate the contribution of the FGMOS to the PSRR+ in the circuit, let us assume only mismatch between the input capacitances. In this case, the PSRR+ predicted for the CMFB is

$$\text{PSRR+}_{\text{CMFB}} \approx \frac{C_{\text{in}}}{C_{\text{T}}} \left[\left(\frac{C_{\text{VDD}}}{C_{\text{T}}} \right)_{f1} - \left(\frac{C_{\text{VDD}}}{C_{\text{T}}} \right)_{f2} \right]^{-1} \tag{4.85}$$

If for example $C_{\text{VDD}} = C_{\text{in}}$ and the mismatch is 1 per cent, the PSRR+$_{\text{CMFB}}$ value would be 40 dB. This value would increase if V_{DD} is also used to generate the reference voltage, in which case the second term in eq. (4.85) would be larger.

Going back to the transconductor circuit, if all these expressions are substituted into eq. (4.79) and again only mismatch between input capacitances is assumed, equal to 1 per cent, the transconductances and input weights are assumed to have the same value, and N2 DC gain is 10, the PSRR+ for the whole circuit would be just 14 dB. This very low value is caused by the tuning amplifier N2, because of the aforementioned reasons. However, if the FGs are not connected together, or the threshold voltage shift is performed without using V_{DD}, the PSRR+ would increase to 36 dB and it would mainly be caused by the connection to V_{DD} in the transconductor FGMOS input transistors[33].

---

[33] In the design example at the end of the chapter the FGs were not connected together.

The same analysis performed for the PSRR− yields

$$\text{PSRR−} \approx G_m R_{out} \left(-R_{outp} g_{mal} \left(1 - \frac{g_{m3r}}{g_{m4r}} a_{1n} \right) + R_{outn} g_{malr} \left(1 - \frac{g_{m3}}{g_{m4}} a_{1p} \right) \right.$$

$$- (A_{CM3p} R_{outp} - A_{CM3n} R_{outn}) - R_{outp} g_{ma3} a_{1p} + R_{outn} g_{ma3r} a_{1n}$$

$$- (A_{CM1p} R_{outp} - A_{CM1n} R_{outn}) \left. \frac{v_{outcm}}{v_{ss}} \right|_{v_{outCM}=0}$$

$$\left. + g_{mal} R_{outp} - g_{malr} R_{outn} \right)^{-1} \tag{4.86}$$

$$A_{CM3p} = \frac{C_{gndf}}{C_T} (g_{mc3} \times R_{1p}) + (g_{mbFc1} \times R_{2p}) \tag{4.87}$$

$$a_{1p} = \frac{(g_{mg1} + g_{dsF1})(g_{mf} R_{outf})_p}{[g_{dsF1} + g_{ds2} + g_{m2}(g_{mf} R_{outf})_p]} + \left. \frac{v_{out(N2)}}{v_{ss}} \right|_{v_{inr}=0} \tag{4.88}$$

$$\left. \frac{v_{outcm}}{v_{ss}} \right|_{v_{outCM}=0}$$

$$\approx \left(1 + \frac{g_{mf4}}{g_{mf3}} \right)$$

$$\times \frac{(g_{mf2} + g_{mbf2})g_{mbFf1} - (g_{mf1} + g_{mbf1})\left(g_{mbFf2} + (C_{gnd}/C_T)g_{mf2}\right)}{(g_{mf1} + g_{mf2} + g_{mbf1} + g_{mbf2})(g_{dsf4} + g_{dsFf2})} \tag{4.89}$$

where C_{gnd}/C_T is the weight associated with the ground connected input in the CMFB.

In this case, the main contribution to the PSRR− comes from the CMFB, due to the connection to ground used to generate the reference voltage. However, as the gain from the common mode feedback nodes to the output is much lower than from the input to the output, the degradation of the PSRR− caused by the FGMOS is not going to be as critical as the degradation of the PSRR+. Hence, for example, under the same assumptions as before and also, considering that the output conductances have the same values and, the same for the transconductances, and the mismatch between both branches capacitances connected to the CMFB is 1 per cent, the PSRR− would be over 36 dB. Improvements can be achieved by generating the reference voltage without using the negative voltage supply, but again this would compromise the power consumption.

4.8 Filter example

This section illustrates the operation of the previously described integrator when it is used as the main building block in a second-order lowpass/bandpass filter. The

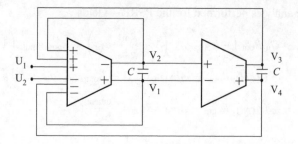

Figure 4.14   Biquad circuit

starting point for the design is the state space equations

$$\dot{x}_1 = -2k_o x_1 - k_o x_2 + k_o u \qquad (4.90)$$

$$\dot{x}_2 = k_o x_1 \qquad (4.91)$$

The lowpass (LP) and bandpass (BP) transfer function are given by X_2 (s) and X_1 (s), respectively. The filter has a fixed quality factor of 0.5 and a cut-off frequency $\omega_0 = k_0$[34]:

$$X_2(s) = \frac{k_o^2}{s^2 + 2k_o s + k_o^2} U(s) \qquad (4.92)$$

$$X_1(s) = \frac{s k_o}{s^2 + 2k_o s + k_o^2} U(s) \qquad (4.93)$$

The block diagram of a circuit that implements the state space equations is shown in Fig. 4.14, where $x_1 = (V_1 - V_2)$, $x_2 = (V_3 - V_4)$ and $u = (U_1 - U_2)$. If the values of the input weights are the same for all inputs, this is $C_i = C_{in}$ for $i = [1, N]$, k_o is given by

$$k_o = \frac{k \beta_1 C_{in} V_b}{C_T C} \qquad (4.94)$$

The filter was designed in a 0.8 μm CMOS technology [214] with nominal threshold voltages of around 0.8 V. The transistor sizes and values of the capacitances used in the design are collected in Table 4.2[35].

A microphotograph of the fabricated circuit is shown in Fig. 4.15. The total area is 0.13 mm^2. Experimental frequency responses for different LP and BP functions are illustrated in Fig. 4.16 and Fig. 4.17. They are obtained for different tuning voltages, V_b, in the range from 20 to 150 mV. The THD measured for different input levels is illustrated from Fig. 4.18 to Fig. 4.21. The maximum signal level is not determined

---

[34] The quality factor and gain could have been made programmable as well just by choosing different coefficients in the state space equations.

[35] Variables refer to Fig. 4.3.

Table 4.2	Design sizes in μm and capacitances values
M_1	4.5/2.5
M_2	4.5/3
M_{a3}	2.5/14
M_{c1}	5/3
M_{c3}	7/2
C_{in}	66 fF
C	5 pF

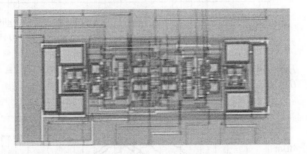

Figure 4.15   Microphotograph of the biquad prototype

by the nonlinearities in the input devices but by output swing limitations. Still the LP output exhibits less than 0.4 per cent distortion for $2V$pp differential input signal at 1.5 V voltage supply and as predicted it is mainly caused by the second-order harmonic. Figure 4.22 shows experimental results for a two-tone test that measures the IM3 in the BP filter. Figure 4.23 illustrates the performance of the CMFB showing how the common mode level remains at 0.75 V. The transient response of the common mode at the output is shown in Figure 4.24 when a $1V$pp differential input signal is applied. It can be seen how the CMRR remains below −40 dB. A summary of the filter performance is given in Table 4.3.

4.9   Summary and conclusions

The FGMOS transistor biased in the strong inversion ohmic region can be used to perform LV linear voltage to current conversion. The advantages of using FGMOS devices in this type of topologies are: a) The effective threshold voltage can be reduced and hence the valid input signal range can be increased since the transistor remains in the strong inversion region even for low/high (depending on whether it is an n- or a p-channel device) values of the input signal; b) The effective input signal at the FG is

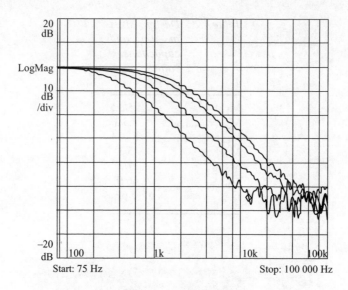

Figure 4.16   Experimental LP filter frequency response $(V_b(V) = 0.02, 0.07, 0.1, 0.15)$

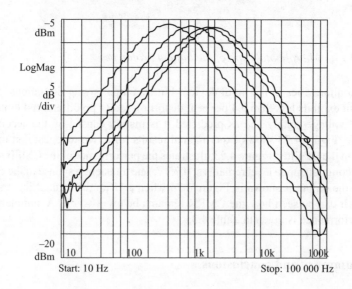

Figure 4.17   Experimental BP filter frequency response $(V_b(V) = 0.02, 0.07, 0.1, 0.15)$

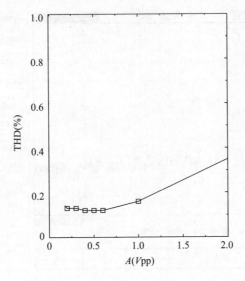

Figure 4.18   THD for a LP function with $f_o = 1.4\ kHz$ and a 500 Hz sinusoidal input

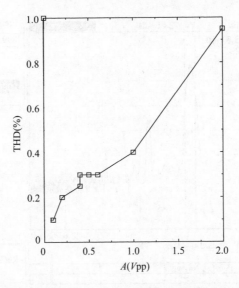

Figure 4.19   THD for a BP function with $f_o = 2\ kHz$ and a 2 kHz sinusoidal input

scaled down by the input weight, which translates into a lower distortion and hence a larger input range; c) The gate to drain coupling in the input transistor contributes to the reduction of the THD since it appears subtracted from the coefficient of the quadratic term in the transistor current law; d) The output resistance can be increased by using two-input FGMOS cascode transistors, which can also simultaneously be used in the common mode feedback strategy and e) Multiple input FGMOS can be

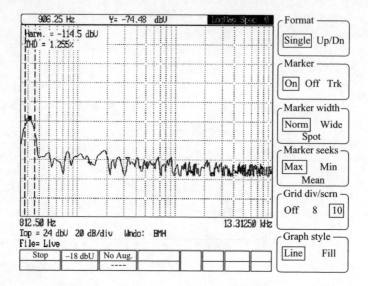

Figure 4.20   THD for a sinusoidal input of 15 mV$_{pp}$ in a LP filter with $f_0 = 2$ kHz

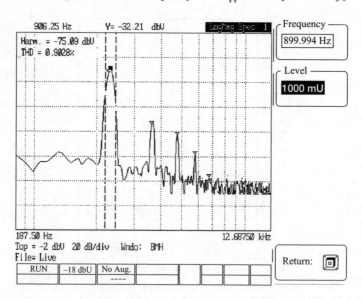

Figure 4.21   THD for a sinusoidal input of 1 V in a LP filter with $f_0 = 2$ kHz

used to perform signal averaging within a single transistor in a CMFB. This results in a very compact, lower power and a more linear topology (when compared with other existing ones).

But also, the transistor presents several drawbacks that the designer should bear in mind. These are as follows: a) As the number of inputs, N, increases, the equivalent

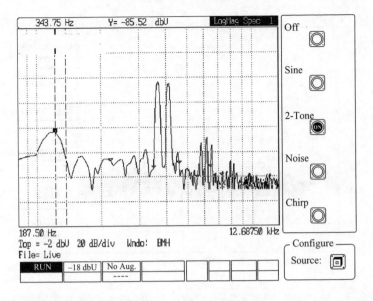

Figure 4.22   Spectrum at the output for a two harmonics input in a BP filter with $f_0 = 2\ kHz$

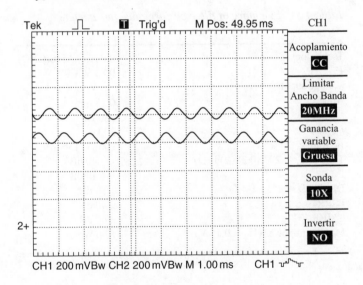

Figure 4.23   Single outputs transient response

transconductance for each input decreases as approximately $1/N$ for the same level of power; b) The DR does not improve much since the equivalent input noise is scaled up by the inverse of the input weight; c) Using the voltage supplies to shift effective threshold voltages can seriously degrade the PSRR; d) The mismatch between the

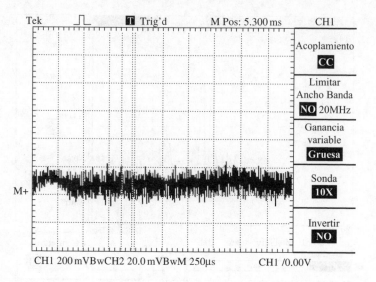

Figure 4.24   Common mode signal sensed at the output for a differential input of 1 Vpp

Table 4.3   Summary of filter performance parameters

Technology	0.8 μm, AMS(CXQ)
V_{Tn}, V_{Tp}	0.82 V, −0.8 V
Voltage supply	1.5 V
Area	0.13 mm^2
f_0, min	380 Hz
f_0, max	2 kHz
V_b Range	20 mV, 150 mV
THD ($V_{pp} < 2$ V, (0.38–2) kHz)	< -40 dB
IM3	< -40 dB
DR (THD < 1%)	> 50 dB
PSRR	> 40 dB
CMRR	> 40 dB
Maximum power ($V_b = 0.15$ V)	13 μW

input capacitances is going to add extra components to the common mode gain and the gain from the power supplies, which will degrade the CMRR and PSRR and e) There is a limitation for the maximum transconductance that can be achieved with these topologies, imposed by the maximum drain voltage that keeps the transistor in the ohmic region, and the maximum width of the device. The reason why there is a maximum value for the width above which there is no benefit on increasing it further is because, a larger width also means larger parasitic capacitances. This would reduce the input weights (because C_T increases) which in its turn would cause the transconductance to decrease, apart from having other undesired effects in the frequency response. An initial solution for this could be to increase the value of the input capacitances, but this would add extra area and would also have a loading effect when the transconductor is used in higher order configurations that would limit the maximum speed. Therefore, this kind of topology is only recommended for low-medium speed applications.

Notation

β_{1i}	In analysis of mismatch ideal β_1 in absence of mismatch
β_i	β parameter for transistor Mi
β_n	β parameter for an n-channel transistor (eq. (4.2))
ε	See eqs. (4.26) and (4.27)
θ	Mobility parameter
ω_o	Filter cut-off frequency (rad/s)
A	Amplitude of a sinusoidal input signal
A_o	N2 DC gain
a_o	Gain from $v_{out(N2)}$ to v_R (eq. (4.38))
a'_{0n}	Parameter equivalent to a'_{0p} in the n side of the integrator
a'_{0p}	See eq. (4.82)
A_1	See eq. (4.34)
a_{1n}	Equivalent to a_{1p} but in the n side of the transconductor
a_{1p}	See eq. (4.88)
A_{CM}	A_{CMp} and A_{CMn} in the absence of mismatch
A_{CMn}	Equivalent to A_{CMp} but in the n side of the transconductor
A_{CMp}	See eq. (4.50)
A_{CM1n}	Equivalent to A_{CM1p} but in the n side of the transconductor
A_{CM1p}	See eq. (4.50)
A_{CM2n}	Equivalent to A_{CM2p} but in the n side of the transconductor
A_{CM2p}	See eq. (4.81)
A_{CM3n}	Equivalent to A_{CM3p} but in the n side of the transconductor
A_{CM3p}	See eq. (4.87)
A_V	N2 small signal gain
B	See eq. (4.9)
B'	See eq. (4.57)
B_1	See eq. (4.35)
B'_4	See eq. (4.68)
C	Integrating capacitance
c_0	See eq. (4.60)
C_{02}	See eq. (4.30)
C_{22}	See eq. (4.31)
C_{42}	See eq. (4.32)
c_1	See eq. (4.61)

C_1'	See eq. (4.18)
$(C_1'/C_T)_n$	(C_1'/C_T) in the n branch of the fully differential integrator (Fig. 4.3) when mismatch is assumed
$(C_1'/C_T)_p$	(C_1'/C_T) in the p branch of the fully differential integrator (Fig. 4.3) when mismatch is assumed
c_2	See eq. (4.62)
C_2'	See eq. (4.19)
C_3'	See eq. (4.20)
C_{CMn}	Capacitance connected to V_{outcm} in n-channel cascode transistors (Fig. 4.3)
$(C_{CMn}/C_T)_n$	(C_{CMn}/C_T) in the n branch of the fully differential integrator (Fig. 4.3) when mismatch is assumed
$(C_{CMp}/C_T + C_{gndf}/C_T)_n$	$(C_{CMp}/C_T + C_{gndf}/C_T)$ in the n branch of the fully differential integrator (Fig. 4.3) when mismatch is assumed
C_{CMp}	Capacitance connected to V_{outcm} in p-channel cascode transistors (Fig. 4.3)
$(C_{CMn}/C_T)_p$	(C_{CMn}/C_T) in the p branch of the fully differential integrator (Fig. 4.3) when mismatch is assumed
$(C_{CMp}/C_T + C_{gndf}/C_T)_p$	$(C_{CMp}/C_T + C_{gndf}/C_T)$ in the p branch of the fully differential integrator (Fig. 4.3) when mismatch is assumed
C_{DBi}	Drain to bulk coupling capacitance in transistor Mi
C_{eq}	See eq. (4.25)
C_{GDi}	Gate to drain coupling capacitance in transistor Mi
C_{gnd}	Value of the input capacitance connected to ground in the CMFB
C_{gndf}	Capacitance connected to ground in p-channel cascode transistors (Fig. 4.3)
C_i for $i = [1, N]$	M1 input capacitances (N1; Fig. 4.1)
C_{in}	Effective input capacitance when only one effective input is considered in N1
C_{inn}	Effective input capacitance in the n branch when mismatch is assumed (Fig. 4.10)
$(C_{in}/C_T)_n$	Effective input weight in the n branch when mismatch is assumed (Fig. 4.3, Fig. 4.10)
C_{inp}	Effective input capacitance in the p branch when mismatch is assumed (Fig. 4.10)
$(C_{in}/C_T)_p$	Effective input weight in the p branch when mismatch is assumed (Fig. 4.3, Fig. 4.10)
C_L	Capacitive load at the output of N2
C_{LN2}	Capacitive load in N1 due to the connection with the tuning amplifier N2
C_R	Capacitance connected to V_R in FGMOS transistors Mn in N2 (Fig. 4.2)

C_T	Total capacitance for all the FGMOS transistors unless the opposite is said
C_{VDD}	Capacitance connected to V_{DD} in the transconductor FGMOS input transistors and also in Mn in N2 and in the CMFB (Fig. 4.1, Fig. 4.2)
$(C_{VDD}/C_T)_{fi}$ for $i = [1,2]$	(C_{VDD}/C_T) when mismatch is considered in the CMFB
C_{VDDf}	Capacitance connected to V_{DD} in n-channel cascode transistors (Fig. 4.3)
$(C_{VDDf}/C_T)_n$	(C_{VDDf}/C_T) in the n branch of the fully differential integrator when mismatch is taken into account
$(C_{VDDf}/C_T)_p$	(C_{VDDf}/C_T) in the p branch of the fully differential integrator when mismatch is taken into account
D	See eq. (4.8)
d_0	See eq. (4.63)
D_1	See eq. (4.36)
d_1	See eq. (4.64)
d_2	See eq. (4.66)
d_3	See eq. (4.69)
f_o	Filter cut-off frequency (Hz)
f_{omax}	Maximum value of the cut-off frequency in the tuning range
f_{omin}	Minimum value of the cut-off frequency in the tuning range
g_{dsbias}	Output conductance of the bias current source in N2
$g_{dsbiasc}$	Output conductance of the bias current source in the CMFB
g_{dsi}	Transistor Mi output conductance
g_{dsFi}	Drain conductance for FGMOS transistor Mi
g_{mbFi}	Bulk transconductance for FGMOS transistor Mi
g_{mc1}^{CM}	See eq. (4.78)
g_{mc1}'	See eq. (4.78)
g_{mc3}^{CM}	See eq. (4.78)
g_{mc3}'	See eq. (4.78)
g_{mgi}	See eq. (4.78)
g_{mf}	N2 transconductance (eq. (4.6))
$(g_{mf}R_{outf})_n$	Gain of the N2 block in the n branch when mismatch is assumed
$(g_{mf}R_{outf})_p$	Gain of the N2 block in the p branch when mismatch is assumed
G_{mi}	Transconductance associated to effective input i (eq. (4.4))
g_{mi}	Transistor Mi transconductance
g_{mi}'	See eq. (4.78)
g_{msfi} for $i = [1,2]$	Fig. 4.13
G_{outn}	Output conductance for the n branch in the fully differential transconductor
G_{outp}	Output conductance for the p branch in the fully differential transconductor (Fig. 4.3) when mismatch is assumed

$H_{BP}(\omega_o)$	Bandpass filter gain at the cut-off frequency
$H_{LP}(0)$	Lowpass filter DC gain
I_{B2}	Bias current in the tuning amplifier (Fig. 4.2)
I_D	M1 drain current (N1. Fig. 4.1)
I_{out}	Fully differential integrator output current (Fig. 4.3)
i_{outi}	Ideal fully differential integrator small signal output current when N2 gain is infinity (eq. (4.26))
i_{outr}	Fully differential integrator small signal output current when N2 gain is not ideal (eq. (4.26))
k	Scaling factor for current mirrors in Fig. 4.3
k_o	Coefficient of the state variables and inputs in eq. (4.90). Given in eq. (4.94)
k_4	See eq. (4.58)
k_5	See eq. (4.65)
k_6	See eq. (4.65)
k_7	See eq. (4.67)
N	Number of inputs in an FGMOS transistor (N1 block; Fig. 4.1)
p	N2 pole
p_1	Transmission pole from $v_{out(N2)}$ to v_R (eq. (4.39))
Q	Filter quality factor
R_1	R_{1n} and R_{1p} when no mismatch is assumed
R_{1n}	Parameter equivalent to R_{1p} for the n side of the transconductor
R_{1p}	See eq. (4.49)
R_2	R_{2n} and R_{2p} in the absence of mismatch
R_{2n}	Parameter equivalent to R_{2p} for the n side of the transconductor
R_{2p}	See eq. (4.49)
R_{out}	Output resistance in the fully differential transconductor (Fig. 4.3) when no mismatch is assumed
R_{outf}	N2 output resistance (eq. (4.6))
R_{outn}	Output resistance for the n branch in the fully differential transconductor (Fig. 4.3) when mismatch is assumed
R_{outp}	Output resistance for the p branch in the fully differential transconductor (Fig. 4.3) when mismatch is assumed
$s_i' = \sqrt{s_i^2}$	
s_i	Power spectral density of noise
u	Filter input (eq. (4.90). ($u_1 - u_2$))
U_1	Single voltage related to u
U_2	Single voltage related to u
V_1	Single voltage related to state variable x_1
V_2	Single voltage related to state variable x_1
V_3	Single voltage related to state variable x_2
V_4	Single voltage related to state variable x_2
V_b	Reference voltage in N2 (Fig. 4.1)
V_D	M1 drain voltage (Fig. 4.2)
V_{D3}	M3 drain voltage (Fig. 4.3)

V_{D3r}	M3r drain voltage (Fig. 4.3)
V_{FG}	M1 FG voltage
V_{FGcn}	FG voltage for n-channel cascode transistor in the integrator in Fig. 4.3 (eq. (4.12))
V_{FGcp}	FG voltage for p-channel cascode transistor in the integrator in Fig. 4.3 (eq. (4.13))
V_i for $i = [1, N]$	FGMOS transistor voltage inputs in N1 (Fig. 4.1)
V_{in}	Voltage at N1 effective input when only one input is considered
V_{ind}	Differential input in the transconductor in Fig. 4.3 when only one effective input is considered in each input FGMOS transistor
V_{inCM}	Common mode at the input (when a single input is considered)
V_{inr}	M1 drain voltage (N1. Fig. 4.1)
V_{jn} for $j = [1, N]$	M1r effective inputs – branch n (Fig. 4.3)
V_{jp} for $j = [1, N]$	M1 effective inputs – branch p (Fig. 4.3)
V_{out}	Differential output for the transconductor in Fig. 4.3
V_{out1}	Mf1 drain voltage (CMFB)
$V_{out(N2)}$	N2 output (Fig. 4.1)
$V_{out(N2)Q}$	Value of $V_{out(N2)}$ when $V_{inr} = V_b$ (eq. (4.6))
V_{outCM}	Common mode at the output
V_{outcm}	Output of the CMFB (Fig. 4.4)
V_{outn}	Single output in the fully differential multiple input integrator (Fig. 4.3)
V_{outp}	Single output in the fully differential multiple input integrator (Fig. 4.3)
V_R	Drain voltage of the input transistor in the tuning amplifier N2 (Fig. 4.2)
V_S	Variable used to refer to the source voltage in the differential pairs (N2 and CMFB)
V_{s1n}	Mc1r source voltage (Fig. 4.3)
V_{s1p}	Mc1 source voltage (Fig. 4.3)
V_{s3n}	Mc3r source voltage (Fig. 4.3)
V_{s3p}	Mc3 source voltage (Fig. 4.3)
V_{Tn}	n-channel transistors threshold voltage. The same is assumed for all the devices unless the opposite is said
V'_{Tn}	Effective threshold voltages for n-channel cascode transistor (eq. (4.16))
V'_{Tp}	Effective threshold voltages for p-channel cascode transistor (eq. (4.17))
$|V_{TP}|$	p-channel transistors threshold voltage (the same for all the devices unless the opposite is said)
x_1	State variable related to bandpass output in eq. (4.90)
x_2	State variable related to lowpass output in eq. (4.90)
z	N2 zero
z_1	Zero in the transfer function between $v_{out(N2)}$ and v_R (eq. (4.40))

Chapter 5

Low power analog continuous-time filtering based on the FGMOS in the strong inversion saturation region

5.1 Introduction

This chapter presents filter designs that use the characteristic features of the FGMOS transistor operating in the strong inversion saturation region to be able to operate at a very low power supply voltage with also a very low power consumption. Again, the implementation of the mathematical functions is greatly improved thanks to the extra degrees of freedom that the device provides, which results in a fewer number of power consuming elements. Besides, the new effective threshold voltages make possible to operate with signal levels that would have switched normal MOS transistors off, hence enabling the reduction of the power supply voltage, while still keeping the signal ranges.

The chapter also analyses the drawbacks of using FGMOS transistors and based on this analysis advises the reader on when not to use them. Thus, for example it shows how in certain cases, the PSRR can be seriously degraded, and if it is a design constraint the transistor should be used in a different way, or even avoided.

Two different integrator designs illustrate the transistor performance. The first one is an audio range linearised G_m-C block in which the linearisation is obtained by combining the quadratic law in different FGMOS devices working in the strong inversion saturation region. The second integrator is again a G_m-C structure but operating in an intermediate frequency (IF) range under a very restrictive power supply voltage constraint.

Finally, the chapter also describes what happens when the input capacitances in the FGMOS transistors are realised in metal/poly instead of poly2/poly1, using as an example the second integrator/filter block. The purpose is to illustrate what would occur in those cases in which only purely digital technologies are available.

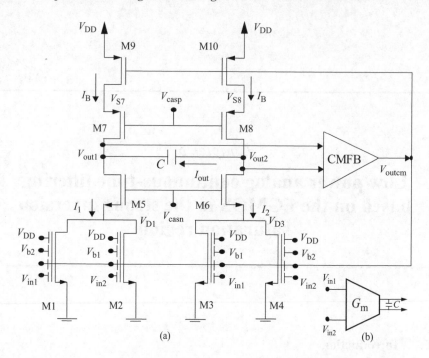

Figure 5.1   (a) Schematic of the proposed FGMOS-based integrator. (b) Symbol

5.2   A G_m-C integrator based on the FGMOS in the strong inversion saturation region for audio applications

This section describes an easy way to perform voltage to current linear conversion based on the operation of the FGMOS transistor in the strong inversion saturation region. It also explains how to use the transistor to maximise the input signal range under low voltage constraints.

Figure 5.1 shows the schematic of a linear G_m-C integrator that uses FGMOS transistors at the input to achieve the linear operation. The circuit works as follows: the input stage is a pseudodifferential amplifier comprised of four equally sized transistors, M1–M4, working as two cross-coupled pairs. Transistors M5–M10 act as active load[36]. A different combination of one input signal and one bias signal is connected to each input transistor in a way that both inputs and both bias are connected to both branches. The aspect ratios and input capacitors are the same for all of them. The bias voltages V_{b1} and V_{b2} are different. A third input is used as part of the common mode feedback mechanism. The fourth input is connected to the maximum available voltage, V_{DD}, in order to reduce the effective threshold voltage according to

---

[36] The block labelled CMFB represents the circuit for the output common mode control, which will be described later on.

eq. (3.3). When the transistors are in the saturation region, the output current (I_{out}) is given by

$$I_{out} = I_1 - I_B = I_B - I_2 = \frac{I_1 - I_2}{2} \tag{5.1}$$

$$I_B = \frac{I_1 + I_2}{2} \tag{5.2}$$

Assuming that transistors M1–M4 are biased in the strong inversion saturation region, and using eq. (2.9), eq. (5.1) can be rewritten as a function of the input voltages:

$$I_{out} = \frac{I_1 - I_2}{2} = \frac{\beta_1 C_{in} C_c (V_{b2} - V_{b1})}{2 C_T^2} \cdot (V_{in1} - V_{in2}) \tag{5.3}$$

where β_1 is the input transistors β parameter[37], C_c is the capacitance connected to either V_{b1} or V_{b2} (depending on the transistor) and C_{in} is the capacitance connected to either V_{in1} or V_{in2}. Equation (5.3) is the function of a linear V/I converter whose output is the current I_{out}, and the input is the differential signal ($V_{in1} - V_{in2}$). The transconductance is given by

$$G_m = \frac{\beta_1 C_{in} C_c (V_{b2} - V_{b1})}{2 C_T^2} \tag{5.4}$$

G_m can be tuned with voltages V_{b2} and V_{b1}. Equation (5.4) shows how values of negative transconductances can also be obtained if necessary. Input capacitances and bias voltages values have to be selected according to eq. (2.9) in order to maintain the MOS transistors working in the strong inversion saturation region.

One of the advantages of the topology in Fig. 5.1 is that the transconductance is completely independent of either the common mode signal or the threshold voltage or the current I_B. Equation (5.4) is valid whenever the input transistors are in the strong inversion saturation region and the second-order effects can be neglected in the quadratic current law. Hence, the effective threshold voltage can be shifted to whichever value is convenient to keep the input transistors in strong inversion by means of the extra input connected to V_{DD}. Besides, the bias current does not need to be constant, which means that it can be adapted to control the common mode at the output as it will be explained later on. In the schematic in Fig. 5.1, I_B is generated by a voltage signal coming from the CMFB. This signal is also connected to one of the inputs in the FGMOS transistors. In this way, it appears as a common mode signal that does not have any effect on the transconductance but however helps to keep the transistors in the right operating region. Hence, if, for example, the bias voltages, V_{b1} and V_{b2} increase during the tuning phase, the constraint for the strong inversion saturation becomes more restrictive. The common mode at the output would tend to decrease in order to balance out the currents between the top and bottom part of the circuit, but the CMFB will correct this by reducing V_{outcm} and hence increasing I_B.

---

[37] From now on β_i corresponds to the β parameter for transistor Mi.

In this way, the positive variation of V_{b1}/V_{b2} will be compensated with a negative variation of V_{outcm}.

5.2.1   Design trade-offs. Power consumption, voltage supply and limits for the transconductance

This section analyses the design trade-offs between power consumption, voltage supply and speed in the transconductor in Fig. 5.1. The speed is proportional to the value of the transconductance. Therefore, the trade-offs for speed are equivalent to the trade-offs for the latter. The maximum value for the transconductance is given by

$$G_{m(max)} = \frac{\beta_1 C_{in} C_c (V_{b2max} - V_{b1min})}{2 C_T^2} \tag{5.5}$$

And the minimum for a given aspect ratio depends on how small the difference $(V_{b2} - V_{b1})$ can be. Nevertheless, the maximum will be the most limiting value.

The maximum and minimum values for the power consumption are a function of the current I_B flowing through transistors M7–M10. I_B is given by[38]

$$I_B = \frac{\beta_{10}}{2}(V_{DD} - V_{outcm} - |V_{Tp}|)^2 \tag{5.6}$$

Its maximum and minimum values can be written as functions of the maximum and minimum values of the voltages at the inputs of the FGMOS transistors M1–M4, and the maximum and minimum value of the voltage at the output of the CMFB:

$$\frac{I_{Bmax}}{2} = \frac{\beta_1}{2}\left(\frac{C_{CM}}{C_T}V_{outcm(min)} + \frac{C_{in}}{C_T}(V_{DD} - 0.2) + \frac{C_c}{C_T}V_b - V_{TFG}\right)^2$$

$$= \frac{\beta_{10}}{4}(V_{DD} - V_{outcm(min)} - |V_{Tp}|)^2 \tag{5.7}$$

$$\frac{I_{Bmin}}{2} = \frac{\beta_1}{2}\left(\frac{C_{CM}}{C_T}V_{outcm(max)} + \frac{0.2C_{in}}{C_T} + \frac{C_c}{C_T}V_b - V_{TFG}\right)^2$$

$$= \frac{\beta_{10}}{4}(V_{DD} - V_{outcm(max)} - |V_{Tp}|)^2 \tag{5.8}$$

where

$$V_{TFG} = V_{Tn} - \frac{C_{VDD}}{C_T}V_{DD} \tag{5.9}$$

$$V_{b1} = V_{b2} = V_b \tag{5.10}$$

C_{VDD} and C_{CM} are the value of the input capacitances which are connected to V_{DD} and V_{outcm}, respectively. Equation (5.7) has been obtained assuming that the maximum

---

[38] From now on $|V_{Tp}|$ will refer to the threshold voltage of p-channel devices. The same value will be assumed for all the devices unless the opposite is said. The same applies for V_{Tn} and n-channel transistors.

signal range is limited by the output swing (which is justified, as in general the transconductor is not going to be used on its own), and the minimum V_{DS} required by the p-channel devices is 0.1 V. Also, transistors M9–M10 determine the I_B current; therefore, its maximum value will be sourced when V_{outcm} is minimum ($V_{outcm(min)}$) and the minimum when V_{outcm} is maximum ($V_{outcm(max)}$). From eq. (5.7) and (5.8), the limits for V_{outcm} are given by

$$V_{outcm(min)} = \frac{1}{C_{CM}/C_T + \sqrt{\beta_{10}/2\beta_1}}$$

$$\times \left(\sqrt{\frac{\beta_{10}}{2\beta_1}} (V_{DD} - |V_{Tp}|) - \frac{C_{in}}{C_T} (V_{DD} - 0.2) - \frac{C_c}{C_T} V_b + V_{TFG} \right) \tag{5.11}$$

$$V_{outcm(max)} = \frac{1}{C_{CM}/C_T + \sqrt{\beta_{10}/2\beta_1}}$$

$$\times \left(\sqrt{\frac{\beta_{10}}{2\beta_1}} (V_{DD} - |V_{Tp}|) - \frac{0.2C_{in}}{C_T} - \frac{C_c}{C_T} V_b + V_{TFG} \right) \tag{5.12}$$

And the maximum and minimum power:

$$P_{limits} = \begin{cases} P_{min} = \beta_{10} V_{DD} (V_{DD} - V_{outcm(max)} - |V_{Tp}|)^2 \\ P_{max} = \beta_{10} V_{DD} (V_{DD} - V_{outcm(min)} - |V_{Tp}|)^2 \end{cases} \tag{5.13}$$

The limits for the power are represented in Fig. 5.2. The vertical plane represents the maximum available tuning voltage (V_{DD}). Hence, for example, for $V_b = 1.25$ V and $V_{DD} = 2$ V, the maximum power consumption required for the circuit to operate

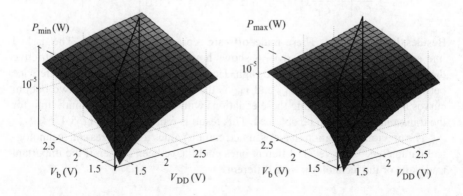

Figure 5.2   Limits for the power established by the tuning voltage and the voltage supply for an input range from 0.2 V to ($V_{DD} - 0.2$) V

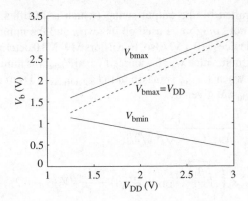

Figure 5.3 V_b limits versus the voltage supply for the maximum and minimum bias current

within the whole tuning range would be 7 μW, whereas the minimum power it would need for the minimum tuning signal value would be 1.8 μW.

Once the limits of power required by the circuit in order to bias the devices properly have been established, the maximum value of the transconductance can also be obtained as follows: equations (5.11) and (5.12) are only valid when transistors M9–M10 are in the strong inversion saturation region and also $V_{outcm} \geq 0$. The maximum and minimum values of V_b that satisfy this are

$$V_{bmax} \leq \frac{C_T}{C_c}\left[-\frac{C_{in}}{C_T}(V_{DD}-0.2)+V_{TFG}+\sqrt{\frac{\beta_{10}}{2\beta_1}}(V_{DD}-|V_{Tp}|)\right] \qquad (5.14)$$

$$V_{bmin} \geq \frac{C_T}{C_c}\left[-\frac{C_{CM}}{C_T}(V_{DD}-|V_{Tp}|)-\frac{0.2C_{in}}{C_T}+V_{TFG}\right] \qquad (5.15)$$

Besides $V_{bmax} \leq V_{DD}$. These trade–offs are represented in Fig. 5.3. The dashed line corresponding to $V_{bmax} \leq V_{DD}$ shows how this condition is more restrictive than eq. (5.20) for this particular design. Hence, the maximum possible difference between the tuning voltages V_{b1} and V_{b2} is the difference between the two bottom curves in Fig. 5.3. Combining these equations with eq. (5.5) the maximum limit for the transconductance can be obtained. This result is represented in Fig. 5.4.

The previous derivations have assumed certain values for the sizes and technological parameters[39]. However, different ones could have been chosen if the important trade-offs or the technology were different.

---

[39] Those corresponding to the design example at the end of section.

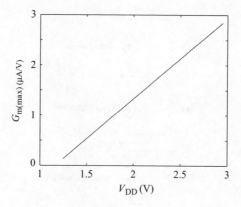

Figure 5.4 $G_{m(max)}$ *versus power supply*

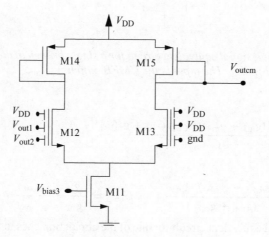

Figure 5.5 *Common mode feedback circuit (CMFB)*

5.2.2 The Common-Mode Feedback Circuit (CMFB): effects on the OTA performance

The functionality of the integrator in Fig. 5.1 is based on its fully balanced configuration and as such will require a CMFB to stabilise the common mode. A CMFB that performs well with this kind of structure is shown in Fig. 5.5. It is based on the same principle as the CMFB described in the previous chapter. The difference between the two of them is the diode connected p-channel transistors used in Fig. 5.5 and therefore the lower gain of the latter.

The output of the CMFB (V_{outcm}) is simultaneously fed back to transistors M1–M4 and M9–M10. Hence, the DC gain from v_{outcm} to one of the single

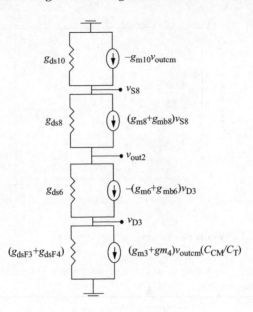

Figure 5.6   Small signal equivalent circuit for a single branch in the transconductor of Fig. 5.1. The input is the CMFB output, v_{outcm}

outputs is

$$\frac{v_{\text{out2}}}{v_{\text{outcm}}} = \frac{[(g_{m3} + g_{m4})(C_{\text{CM}}/C_{\text{T}}) + g_{m10}]}{G_{\text{outn}} + G_{\text{outp}}} \tag{5.16}$$

with

$$G_{\text{outn}} = \frac{(g_{\text{dsF3}} + g_{\text{dsF4}}) \cdot g_{\text{ds6}}}{(g_{m6} + g_{\text{mb6}})}, \qquad G_{\text{outp}} = \frac{g_{\text{ds8}} \cdot g_{\text{ds10}}}{(g_{m8} + g_{\text{mb8}})} \tag{5.17}$$

The small signal equivalent circuit for one of the circuit branches in Fig. 5.1 is shown in Fig. 5.6. The subscripts refer to the name of the corresponding device.

The previous equations have been obtained under the assumptions

$$(g_{m6} + g_{\text{mb6}}) \gg (g_{\text{ds6}} + g_{\text{dsF3}} + g_{\text{dsF4}}) \tag{5.18}$$

$$(g_{m8} + g_{\text{mb8}}) \gg (g_{\text{ds8}} + g_{\text{ds10}}) \tag{5.19}$$

Equation (5.16) shows how the gain is increased by applying feedback in the p- and n-channel transistors. In this way, it is possible to have a low gain CMFB and the variations at the output of the latter can be kept small enough to not exceed the output swing constraint, which becomes very restrictive as the value of the voltage supply is reduced.

The rationale behind this feedback mechanism can also be explained from a large signal point of view: when the tuning voltages (V_{b2}, V_{b1}), or the common mode at the input (V_{inCM}) change, the common mode current does it as well. When this happens the output of the CMFB changes in such a way that the variation of the input

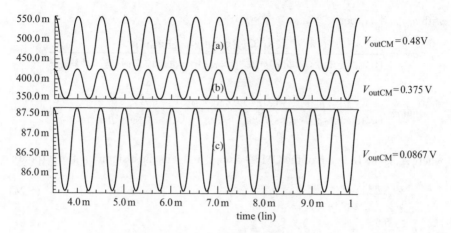

Figure 5.7 *(a) Evolution of the output when a sinusoidal input is applied to the transconductor. Feedback at the load and input transistors. (b) Evolution of the output when a sinusoidal input is applied to the transconductor. Feedback at the load transistors. (c) Evolution of the output when a sinusoidal input is applied to the transconductor. No feedback*

transistors FG common mode operating point is up to a certain extent compensated. Simultaneously, the I_B current changes to correct for the still remaining variation.

Figure 5.7 illustrates this showing how a single output changes when a sinusoidal differential input is applied. The DC level of the output signal represents common mode at the output, V_{outCM}. The reference voltage it is being compared with is 0.5 V. It can be observed how when only the gate of the p-channel transistors is connected to the output of the common mode circuit, the common mode at the output changes more abruptly than when the feedback mechanism is applied to both, the p and the n parts.

Another interesting aspect worth analysing is how this feedback might affect the frequency response of the integrator. If the frequency response of the CMFB is modelled as a single pole function and included into the small signal equations of the integrator, for every single branch

$$sCv_{out2} = -g_{m3}v_{in1} - g_{m4}v_{in2} - \frac{a_{02}}{s+p}(v_{out1} + v_{out2})$$

$$-\left(\frac{(g_{dsF3} + g_{dsF4}) \cdot g_{ds6}}{(g_{m6} + g_{mb6})} + \frac{g_{ds8} \cdot g_{ds10}}{(g_{m8} + g_{mb8})}\right)v_{out2} \qquad (5.20)$$

$$sCv_{out1} = -g_{m1}v_{in1} - g_{m2}v_{in2} - \frac{a_{01}}{s+p}(v_{out1} + v_{out2})$$

$$-\left(\frac{(g_{dsF1} + g_{dsF2}) \cdot g_{ds5}}{(g_{m5} + g_{mb5})} + \frac{g_{ds7} \cdot g_{ds9}}{(g_{m7} + g_{mb7})}\right)v_{out1} \qquad (5.21)$$

with a_{02} and a_{01} being

$$a_{02} = \frac{[(C_{CM}/C_T)(g_{m3}+g_{m4})+g_{m10}]}{((g_{dsF3}+g_{dsF4})\cdot g_{ds6})/(g_{m6}+g_{mb6})+(g_{ds8}\cdot g_{ds10})/(g_{m8}+g_{mb8})}(A_o \cdot p)$$
$$(5.22)$$

$$a_{01} = \frac{[(C_{CM}/C_T)(g_{m1}+g_{m2})+g_{m9}]}{((g_{dsF1}+g_{dsF2})\cdot g_{ds5})/(g_{m5}+g_{mb5})+(g_{ds7}\cdot g_{ds9})/(g_{m7}+g_{mb7})}(A_o \cdot p)$$
$$(5.23)$$

where A_o is the DC gain of the CMFB circuit, and p its most significant root (pole):

$$A_o \approx \frac{g_{m12}(C_{in}/2C_T)}{g_{m15}} \tag{5.24}$$

$$p \approx \frac{g_{m15}}{C_{LF}} \tag{5.25}$$

C_{in} is the value of the input capacitances connected to V_{out1} and V_{out2} in the CMFB[40]. C_{LF} is the load capacitance at the output of the CMFB. Its main contributors are M15 gate to source capacitance, the gate to source capacitances of the bias transistors in the transconductor, M9 and M10, and the input capacitances C_{in} of devices M1–M4.

Assuming that the devices are matched ($a_{01} = a_{02} = a_0$), adding both equations, and taking into consideration that $g_{m3} = g_{m2}$ and $g_{m4} = g_{m1}$, the differential equation for the common mode at the output is obtained:

$$sCv_{outCM} = -(g_{m3} + g_{m1})v_{incM} - \frac{2a_0}{s+p}v_{outCM} - (G_{outn}' + G_{outp}')v_{outCM}$$
$$(5.26)$$

where

$$G_{outn}' = \frac{(g_{dsF3} + g_{dsF4})\cdot g_{ds6}}{(g_{m6} + g_{mb6})} + \frac{(g_{dsF1} + g_{dsF2})\cdot g_{ds5}}{(g_{m5} + g_{mb5})} \tag{5.27}$$

$$G_{outp}' = \frac{g_{ds8}\cdot g_{ds10}}{(g_{m8} + g_{mb8})} + \frac{g_{ds7}\cdot g_{ds9}}{(g_{m7} + g_{mb7})} \tag{5.28}$$

The denominator of the common mode transfer function has the form

$$D(s) = s^2C + s[Cp + (G_{outn}' + G_{outp}')] + [2a_0 + (G_{outn}' + G_{outp}')p] \tag{5.29}$$

The common mode transfer function has then two poles whose location in the real-imaginary plane will depend on the relationship between the different parameters. If the radicand in eq. (5.29) is positive then the system will have two negative poles, being dominant the lowest frequency one. Otherwise, the poles will be complex conjugate and the transient response will depend on the values of the real and imaginary

---

[40] The input capacitance ratio has been assumed to be C_{in}/C_T, although a different value could have been chosen (it does not need to be related to the transconductor input weight).

parts. This will happen whenever

$$8a_0C > [(G_{\text{outn}}' + G_{\text{outp}}') - Cp]^2 \tag{5.30}$$

In this case, the output will show some ringing until it settles which will happen faster as the absolute value of the real term in eq. (5.31) increases:

$$\frac{-[Cp + (G_{\text{outn}}' + G_{\text{outp}}')]}{C} \tag{5.31}$$

The previous two equations show how the higher the bandwidth of the CMFB is the smaller settling time. Also, in order to minimise the ringing the gain of the CMFB should not be too high. That is one of the reasons why a low gain amplifier is chosen for this topology. The gain and bandwidth can still be high enough to correct for common mode variations and simultaneously avoid falling into oscillations.

Summarising, the advantages of using FGMOS transistors in the CMFB are the following:

1. Design simplicity: the common mode at the output can be sensed in a single device. The risk of increasing the THD with the common mode feedback is reduced thanks to the linear summation.
2. The input transistors can be in the strong inversion saturation region for common mode output levels that would not make it possible in normal MOS devices. This has several advantages such as an increased bandwidth and lower offset.
3. The dominant pole can be shifted towards higher frequencies by increasing the bias current in the differential pair. This is possible even for very low values of V_{DD} by reducing the effective threshold voltages of the input transistors with an extra input connected to V_{DD}. In this way, it is possible to minimise the risk of ringing as predicted by eq. (5.30).

5.2.2.1 Common Mode Rejection Ratio (CMRR)

The topology in Fig. 5.1 is completely symmetrical, and because of this ideally the common mode gain at the output should be zero. However, in reality, non-ideal effects such as mismatch will increase the gain [160]. Yet, due to the symmetry of the topology the common mode gain will still be smaller than in a single-ended version. A small signal equivalent circuit for the analysis of the common mode DC gain is shown in Fig. 5.8. The common-mode DC gain is given by

$$\left. \frac{v_{\text{out2}}}{v_{\text{inCM}}} \right|_{v_{\text{ind}}=0} \approx -\left[\left(\frac{C_{\text{in}}}{C_T} \right)_3 g_{\text{m3}} + \left(\frac{C_{\text{in}}}{C_T} \right)_4 g_{\text{m4}} \right]$$

$$\times \left(2A_0 \left[g_{\text{m10}} \frac{(g_{\text{dsF3}} + g_{\text{dsF4}} + g_{\text{m6}} + g_{\text{mb6}})}{g_{\text{m6}} + g_{\text{mb6}}} \right. \right.$$

$$\left. \left. + \left(\frac{C_{\text{CM}}}{C_T} \right)_3 g_{\text{m3}} + \left(\frac{C_{\text{CM}}}{C_T} \right)_4 g_{\text{m4}} \right] \right)^{-1} \tag{5.32}$$

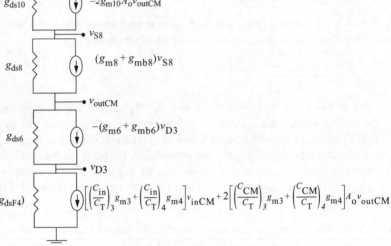

Figure 5.8   Small signal equivalent circuit for the calculation of the common mode output gain

Equation (5.32) represents the gain of one of the single outputs. If both symmetrical branches were identical the common mode gain resulting from subtracting two identical equations from both single branches would be zero. In practice, the mismatch between devices will give rise to two different equations. The expression for the other branch can be obtained from eq. (5.32) just by changing the subscripts to those corresponding to the devices in it. New subscripts have been added to account for the mismatch between ideally identical FGMOS. They refer again to the name of the device they correspond to. From now on, for the sake of simplicity the mismatch between capacitances will be modelled in the form of a parameter Δ in the following way:

$$\left(\frac{C_{in}}{C_T}\right)_1 = \frac{C_{in}}{C_T} \qquad \left(\frac{C_{in}}{C_T}\right)_{i=[2,4]} = \frac{C_{in}}{C_T}(1 + \Delta_{1i}) \qquad (5.33)$$

$$\left(\frac{C_{CM}}{C_T}\right)_1 = \frac{C_{CM}}{C_T} \qquad \left(\frac{C_{CM}}{C_T}\right)_{i=[2,4]} = \frac{C_{CM}}{C_T}(1 + \Delta_{2i}) \qquad (5.34)$$

The differential output gain when the input is a common mode signal, assuming that $g_{dsF3} + g_{dsF4} \ll g_{m6} + g_{mb6}, g_{dsF1} + g_{dsF2} \ll g_{m5} + g_{mb5}$ and only mismatch between the capacitances is[41]

$$\frac{(v_{out1} - v_{out2})}{v_{inCM}}\bigg|_{v_{ind}=0}$$

$$\approx \left([g_{m3}(\Delta_{13} - \Delta_{12}) + g_{m4}\Delta_{14}] \left[g_{m10} + \left(\frac{C_{CM}}{C_T}\right)(g_{m3} + g_{m4}) \right] \right.$$

$$+ \left(\frac{C_{CM}}{C_T}\right)(g_{m3} + g_{m4})[g_{m3}(\Delta_{22} - \Delta_{23}) - g_{m4}\Delta_{24}]\right)$$

$$\times \left(2A_o \left(\frac{C_T}{C_{in}}\right) \left[g_{m10} + \left(\frac{C_{CM}}{C_T}\right)(g_{m3} + g_{m4}) \right] \right.$$

$$\times \left[g_{m10} + \left(\frac{C_{CM}}{C_T}\right)[(g_{m3} + g_{m4}) + (\Delta_{23} + \Delta_{22})g_{m3} + \Delta_{24}g_{m4}] \right] \right)^{-1}$$

(5.35)

Taking into account that

$$G_m = \left(\frac{C_{in}}{C_T}\right)(g_{m4} - g_{m3})$$

(5.36)

and assuming that the transconductor is designed in such a way that

$$g_{m10} \approx (g_{m3} + g_{m4})$$

(5.37)

eq. (5.35) can be rewritten as

$$\frac{(v_{out1} - v_{out2})}{v_{inCM}}\bigg|_{v_{ind}=0}$$

$$\approx \left([(\Delta_{13} - \Delta_{12}) + k_1\Delta_{14}]\left(1 + \frac{C_{CM}}{C_T}\right) + \left(\frac{C_{CM}}{C_T}\right)[\Delta_{22} - \Delta_{23} - k_1\Delta_{24}] \right)$$

$$\times \left(2A\left(\frac{C_T}{C_{in}}\right)\left(1 + \frac{C_{CM}}{C_T}\right)\left[(1+k_1)\left(1 + \frac{C_{CM}}{C_T}\right)\right. \right.$$

$$\left. + \frac{C_{CM}}{C_T}(\Delta_{23} + \Delta_{22} + k_1\Delta_{24})\right] \right)^{-1}$$

(5.38)

$$k_1 = \left(1 + \frac{G_m}{g_{m3}}\frac{C_T}{C_{in}}\right)$$

(5.39)

---

[41] This assumption is made for the sake of clarity to give the reader an idea of how much the input capacitances in the FGMOS devices can degrade the performance in terms of CMRR. The derivation of a more accurate expression for the CMRR from eq. (5.32) is however trivial.

This gain represents the effect of the common mode signals on the OTA differential output. Ideally it should be zero or at least negligible when compared with the differential gain. The latter can be obtained with the same equivalent circuit if the common mode contributions are cancelled and a differential input is considered instead. The result is in eq. (5.40)[42]. The Common Mode Rejection Ratio (CMRR) is given by the ratio between the differential gain when the input signal is differential (eq. (5.40)), and eq. (5.38):

$$\frac{v_{\text{out}}}{v_{\text{ind}}} \approx \frac{G_m}{G_{\text{outn}} + G_{\text{outp}}} \tag{5.40}$$

$$\text{CMRR} \approx -\frac{G_m}{G_{\text{outn}} + G_{\text{outp}}} \times \left(A_o \left(\frac{C_T}{C_{\text{in}}} \right) \left(1 + \frac{C_{\text{CM}}}{C_T} \right) \right.$$

$$\times \left[(1 - k_1) \left(1 + \frac{C_{\text{CM}}}{C_T} \right) + (\Delta_{23} - \Delta_{22} + k_1 \Delta_{24}) \frac{C_{\text{CM}}}{C_T} \right] \right)$$

$$\times \left([(\Delta_{13} - \Delta_{12}) + k_1 \Delta_{14}] \left(1 + \frac{C_{\text{CM}}}{C_T} \right) + \left(\frac{C_{\text{CM}}}{C_T} \right) [\Delta_{22} - \Delta_{23} - k_1 \Delta_{24}] \right)^{-1}$$

$$\tag{5.41}$$

Equation (5.41) shows that:

1. The rejection of the common mode signals will increase as the CMFB gain increases.
2. Increasing the ratio C_{CM}/C_T will also increase the CMRR.
3. The exact value of the CMRR depends on the value of the transconductance, more specifically on the ratio between the latter and the individual transistors' transconductances, which will in their turn be a function of the bias current I_B. Hence, for example, for extreme transconductance values such as $G_m = (-C_{\text{in}/C_T})g_{m3}$, eq. (5.41) can be written as

$$\text{CMRR} \approx \frac{G_m}{G_{\text{outn}} + G_{\text{outp}}}$$

$$\times \frac{A_o(C_T/C_{\text{in}})(1 + C_{\text{CM}}/C_T)[(1 + C_{\text{CM}}/C_T) + (\Delta_{22} + \Delta_{23})(C_{\text{CM}}/C_T)]}{(\Delta_{13} - \Delta_{12})(1 + C_{\text{CM}}/C_T) + (C_{\text{CM}}/C_T)(\Delta_{22} - \Delta_{23})}$$

$$\tag{5.42}$$

In order to illustrate this result, let us assume a typical differential gain of 40 dB, $A_o = 0.5$, capacitances ratios of 0.2, and a maximum deviation of

---

[42] This equation has not considered mismatch between transistors because the latter would not affect the CMRR significantly.

1 per cent. The minimum value of the CMRR predicted by eq. (5.42) would be 82 dB.

Let us now consider the case in which $G_m \approx g_{m3}$ instead. Equation (5.41) would then be

$$\text{CMRR} \approx \frac{G_m}{G_{\text{outn}} + G_{\text{outp}}} \times \left(A_0 \left(\frac{C_T}{C_{\text{in}}} \right) \left(1 + \frac{C_{\text{CM}}}{C_T} \right) \left[\left(-\frac{C_T}{C_{\text{in}}} \right) \left(1 + \frac{C_{\text{CM}}}{C_T} \right) \right.\right.$$

$$+ (\Delta_{23} + \Delta_{22}) \frac{C_{\text{CM}}}{C_T} + \left(\frac{C_T}{C_{\text{in}}} + 1 \right) \frac{C_{\text{CM}}}{C_T} \Delta_{24} \bigg] \bigg)$$

$$\times \left(\left[(\Delta_{13} - \Delta_{12}) + \left(\frac{C_T}{C_{\text{in}}} + 1 \right) \Delta_{14} \right] \left(1 + \frac{C_{\text{CM}}}{C_T} \right) \right.$$

$$+ \left(\frac{C_{\text{CM}}}{C_T} \right) \left[\Delta_{22} - \Delta_{23} - \left(\frac{C_T}{C_{\text{in}}} + 1 \right) \Delta_{24} \right] \bigg)^{-1} \tag{5.43}$$

And the value of the CMRR for the same hypothetical specifications would be 53 dB. Although the CMRR value appears to be much smaller, one of the assumptions is not quite accurate for this particular case: the fact that in both situations the differential gain would be identical. This is only true if the output resistance changes in the same way as the transconductance, which does not happen because the bias current does not follow the transconductance variation trend. Hence, in reality, the differential gain will be higher for larger G_m.

5.2.3   Second-order effects

5.2.3.1   Total harmonic distortion

The functionality of the transconductor, described by eq. (5.3), was derived assuming ideal square functions of V_{GS} for the currents in the input transistors operating in the strong inversion saturation region. This is an approximation though which does not take into account second-order effects such as mobility reduction and channel length modulation that cause deviations from this ideal behaviour. A better fitting function for the mobility would be given by eq. (4.55). The channel-length modulation as a source of nonlinear distortion is negligible in this structure because of its high output resistance. It will be thus not considered in the following analysis.

Replacing eq. (4.55) for the mobility in each of the expressions for the currents in Fig. 5.1 input transistors, and using Maclaurin series expansion, eq. (5.44) is obtained, where $V_{\text{CM}}' = V_{\text{inCM}} + (C_{\text{CM}}/C_{\text{in}})V_{\text{outcm}} + (C_c/C_{\text{in}})V_{b2}$, and V_{inCM} is the common mode at the effective input/output:

$$I_{out} \approx \frac{\beta_1}{2}\left(\frac{C_{in}}{C_T}\right)\left(V_{CM}' - \frac{C_T}{C_{in}}V_{TFG}\right) \times \frac{\theta(V_{CM}' - (C_T/C_{in})V_{TFG}) + 2(C_T/C_{in})}{[\theta(V_{CM}' - (C_T/C_{in})V_{TFG}) + C_T/C_{in}]^2}V_{ind}$$

$$- \frac{\beta_1}{2}\left(\frac{C_{in}}{C_T}\right)\left(V_{CM}' - \frac{C_T}{C_{in}}V_{TFG} + \frac{C_c}{C_{in}}(V_{b1} - V_{b2})\right)$$

$$\times \frac{\theta(V_{CM}' - (C_T/C_{in})V_{TFG} + (C_c/C_{in})(V_{b1} - V_{b2})) + 2(C_T/C_{in})}{[\theta(V_{CM}' - (C_T/C_{in})V_{TFG} + (C_c/C_{in})(V_{b1} - V_{b2})) + C_T/C_{in}]^2}V_{ind}$$

$$+ \left[\frac{(\theta\beta_1/8)(C_T/C_{in})}{[\theta(V_{CM}' - (C_T/C_{in})V_{TFG} + (C_c/C_{in})(V_{b1} - V_{b2})) + C_T/C_{in}]^4}\right.$$

$$\left. - \frac{(\theta\beta_1/8)(C_T/C_{in})}{[\theta(V_{CM}' - (C_T/C_{in})V_{TFG}) + C_T/C_{in}]^4}\right]V_{ind}^3 \qquad (5.44)$$

The THD is hence given by [171][43]

$$\text{THD} = \frac{\theta}{16}\left(\left[\frac{(C_T/C_{in})^2}{[\theta(V_{CM}' - (C_T/C_{in})V_{TFG} + (C_c/C_{in})(V_{b1} - V_{b2})) + C_T/C_{in}]^4}\right.\right.$$

$$\left.\left. - \frac{(C_T/C_{in})^2}{[\theta(V_{CM}' - (C_T/C_{in})V_{TFG}) + C_T/C_{in}]^4}\right]A^2\right)\left(\left(V_{CM}' - \frac{C_T}{C_{in}}V_{TFG}\right)\right.$$

$$\times \frac{\theta(V_{CM}' - (C_T/C_{in})V_{TFG}) + 2(C_T/C_{in})}{[\theta(V_{CM}' - (C_T/C_{in})V_{TFG}) + C_T/C_{in}]^2} - \left(V_{CM}' - \frac{C_T}{C_{in}}V_{TFG} + \frac{C_c}{C_{in}}(V_{b1} - V_{b2})\right)$$

$$\times \left. \frac{\theta(V_{CM}' - (C_T/C_{in})V_{TFG} + (C_c/C_{in})(V_{b1} - V_{b2})) + 2(C_T/C_{in})}{[\theta(V_{CM}' - (C_T/C_{in})V_{TFG} + (C_c/C_{in})(V_{b1} - V_{b2})) + C_T/C_{in}]^2}\right)^{-1} \qquad (5.45)$$

where A is the amplitude of a sinusoidal input signal. Theoretical values of THD for different values of common mode input voltages and $(V_{b1} - V_{b2})$ are shown in Fig. 5.9. The values of the parameters used to obtained these graphs are the same as for the design example in Section 5.2.5.1 and θ is 0.234 V^{-1} [44]. Curves of simulated THD for this design are given in Fig. 5.15. They show how the THD remains low in a wide range of G_m [45]. Therefore, as long as the noise floor is also low in the same range, the circuit will have a good performance in terms of DR in the whole programmability range.

The main conclusion that can be obtained from the theoretical expressions is that having FGMOS transistors at the input reduces distortion because of the compression of the effective input voltage at the gate by a factor of C_{in}/C_T.

[43] Ignoring mismatch between the two symmetrical branches.
[44] This value can be derived from parameters of the BSIM49 model used in the design example in section 5.2.5.1.
[45] This is equivalent to a wide range of the difference $(V_{b1} - V_{b2})$.

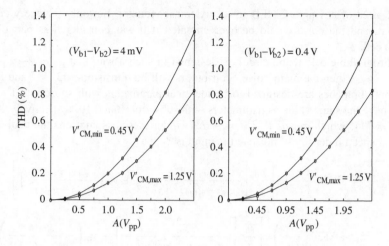

Figure 5.9   Theoretical Total Harmonic Distortion for two different $(V_{b1} - V_{b2})$ values

Table 5.1   Equivalent input noise contribution for every transistor in one of the branches in Fig. 5.1

	v_{in}/S_i' $M3 = M4$	$S_i = S_i'^2$
M6	$\dfrac{[g_{dsF3} + g_{dsF4}](C_T/C_{in})}{(g_{ds6} + g_{dsF3} + g_{dsF4} + g_{m6} + g_{mb6})(g_{m3} - g_{m4})}$	$4KTg_{m6}\left[\dfrac{\rho_n g_{m6}}{W_6 L_6 f} + \gamma_n\right]$
M8	$\dfrac{g_{ds10}(C_T/C_{in})}{(g_{ds8} + g_{ds10} + g_{m8} + g_{mb8})(g_{m3} - g_{m4})}$	$4KTg_{m8}\left[\dfrac{\rho_p g_{m8}}{W_8 L_8 f} + \gamma_p\right]$
Mi $(i = [3,4])$	$\dfrac{(C_T/C_{in})}{(g_{m3} - g_{m4})}$	$4KTg_{m(3,4)}\left[\dfrac{\rho_n g_{m(3,4)}}{W_3 L_3 f} + \gamma_n\right]$
M10	$\dfrac{(C_T/C_{in})}{(g_{m3} - g_{m4})}$	$4KTg_{m10}\left[\dfrac{\rho_p g_{m10}}{W_{10} L_{10} f} + \gamma_p\right]$

5.2.3.2   Noise

Noise of the low voltage transconductor

Approximate expressions for the equivalent noise contribution of each transistor at the effective input are collected in Table 5.1. Similar equations can be obtained for the symmetrical circuit branch just by changing the subscripts. This model does not take into account parasitic capacitance frequency effects, and therefore will not be accurate for high frequency calculations. However, it is good enough to give an idea of the effects caused by the FGMOS transistors. The table does not show the contribution of the CMFB. The latter will be analysed separately since there is no need of having a CMFB for every transconductor, but only for every circuit output, and the noise would be overestimated if it was considered as part of the main cell.

From Table 5.1, if the circuit is designed in such a way that $g_{dsF3} + g_{dsF4} \ll g_{m6} + g_{mb6}$ then the main noise contributors will be transistors M3, M4 and M10. As both branches are designed to be identical parameters with subscripts 1 and 9 will have the same value as parameters with subscripts 3 and 10, respectively. Also, M3 = M4, which means $W_3 L_3 = W_4 L_4$. Under these assumptions the equivalent power spectral density of noise at the input is[46]

$$\frac{\overline{v_{in}^2}}{\Delta f} = 8KT \left(\frac{C_T}{C_{in}} \right)^2 \left[\frac{\rho_n}{f} \left[\frac{g_{m3}^2 + g_{m4}^2}{(g_{m3} - g_{m4})^2 W_3 L_3} \right] + \gamma_n \left(\frac{g_{m3} + g_{m4}}{(g_{m3} - g_{m4})^2} \right) \right]$$

$$+ 8KT \left(\frac{C_T}{C_{in}} \right)^2 \left[\frac{\rho_p}{f} \left[\frac{g_{m10}^2}{(g_{m3} - g_{m4})^2 W_{10} L_{10}} \right] + \gamma_p \left(\frac{g_{m10}}{(g_{m3} - g_{m4})^2} \right) \right]$$

$$(5.46)$$

This expression can be simplified using the parameter k_1 previously defined:

$$\frac{\overline{v_{in}^2}}{\Delta f} = 8KT \left(\frac{C_T}{C_{in}} \right)^2 \left[\frac{1}{f} \left[\frac{(1+k_1)^2}{(1-k_1)^2} \right] \left(\frac{\rho_n}{W_3 L_3} + \frac{\rho_p}{W_{10} L_{10}} \right) \right.$$

$$\left. - \frac{2}{f} \frac{(k_1)}{(1-k_1)^2} \frac{\rho_n}{W_3 L_3} + \frac{(1+k_1)}{(1-k_1)^2} \frac{(\gamma_n + \gamma_p)}{g_{m3}} \right]$$

$$(5.47)$$

Let us now consider as an example the limit condition, $G_m = (-C_{in}/C_T) g_{m3}$. Equation (5.47) can then be rewritten as

$$\frac{\overline{v_{in}^2}}{\Delta f} = 8KT \left(\frac{C_T}{C_{in}} \right)^2 \left[\frac{1}{f} \left(\frac{\rho_n}{W_3 L_3} + \frac{\rho_p}{W_{10} L_{10}} \right) + \frac{\gamma_n + \gamma_p}{g_{m3}} \right]$$

$$(5.48)$$

If the same derivation was carried out for normal MOS devices ($C_{in} \to C_T$) the final expression would be the same, except for the, larger than 1, multiplicative factor $(C_T/C_{in})^2$. In simple terms, this means that the power spectral density of noise at the input increases. But, going back to the linearity analysis performed in the previous section, the power spectral density of the signal for a certain THD requirement also increases. So, the final dynamic range will not degrade. In any case, having the same topology with normal MOS transistors would not be possible since multiple inputs at the gate are needed in order to implement the right functionality. A possible modification would be to have the bias signals connected at the sources of the MOS devices [164], but this would make the final structure much more complex since

---

[46] In the case where the FGMOS devices' gate to drain parasitic capacitances are not small enough for the assumption $g_{dsF3} + g_{dsF4} \ll g_{m6} + g_{mb6}$ to be valid, there would be an extra term in eq. (5.46), that would also have to be considered.

extra circuitry would be needed to generate those voltages. Also, the latter would be connected to low impedance nodes and this would require a considerable amount of extra power for the biasing circuits to be able to cope with the low impedance loads.

The power spectral density of noise at the output can be obtained from eq. (5.46) just by multiplying the input noise by the square of the gain. In this case, the scaling factor given by the effective input weight would disappear and the effect of the FGMOS would be as a reduction of the noise due to the lower output resistance and gain because of the parasitic capacitances.

The general conclusions that can be drawn from eq. (5.46) are:

1. In order to reduce the flicker noise, transistors M1, M2, M3, M4, M9 and M10 should be designed as big as possible. The problem with this is that as the area of the transistors increases the frequency response is poorer. The same happens to the distortion and the DC gain of the transconductor. This could be sorted out by increasing the input capacitances accordingly, but in this case, the area would increase, and also the loading effects when the transconductor is used as part of a bigger system, which would degrade its frequency response.

2. The main advantage of using FGMOS is the small number of devices that are required to linearise the structure which, in terms of noise, is equivalent to having less noise contributors.

Common mode feedback circuit noise
As it was mentioned before the noise analysis for the CMFB deserves a separate section since only one of these blocks is required for every couple of output nodes.

Table 5.2 Equivalent input noise contribution for every transistor in the CMFB

	v_{in}/S'_i
M12	$\left(\dfrac{C_T}{C_{in}}\right)\dfrac{1}{g_{m12}}$
M13	$\cong \left(\dfrac{C_T}{C_{in}}\right)\dfrac{1}{g_{m12}}$
M14	$\left(\dfrac{C_T}{C_{in}}\right)\dfrac{1}{g_{m12}}$
M15	$\cong \left(\dfrac{C_T}{C_{in}}\right)\dfrac{1}{g_{m12}}$

Table 5.2 summarises the approximate equivalent noise at the input of the CMFB for every transistor. The subscripts refer to Fig. 5.5. The input capacitance ratio has been assumed to be C_{in}/C_T, although a different value could have been chosen. (It does not need to be related to the transconductor input weight.)

And the equivalent power spectral noise density at the input for the CMFB is

$$\frac{\overline{\Delta v_{\text{out1}}^2}}{\Delta f} \approx 4KT \left(\frac{C_T}{C_{in}}\right)^2$$

$$\times \left[\frac{1}{f}\left(\frac{2\rho_n}{W_{12}L_{12}} + 2\left(\frac{g_{m14}}{g_{m12}}\right)^2 \frac{\rho_p}{W_{14}L_{14}}\right) + 2\left(\frac{\gamma_n}{g_{m12}} + \frac{g_{m14}\gamma_p}{g_{m12}^2}\right)\right] \tag{5.49}$$

Deviations in transconductances and sizes between similar transistors have been neglected.

The analysis of this block draws similar conclusions to those obtained for the transconductor circuit. Again, the flicker noise can be reduced by increasing the transistors areas, whereas the thermal noise is inversely proportional to the transconductance which on the other hand has to have a minimum value to keep an adequate bandwidth.

The noise in eq. (5.50) will not have a significant effect in the performance of the whole transconductor block. The reason for this is that it is a common mode signal and thus its effect will be almost the same for both symmetrical branches. Its exact contribution will depend on the common mode rejection ratio but, as a first approach, and for the values of CMRR previously obtained it can be considered negligible. Ideally, in absence of mismatch, it would be zero:

$$\frac{\overline{\Delta v_{\text{outcm}}^2}}{\Delta f} \approx (KT)\left[\frac{1}{f}\left(\frac{2\rho_n}{W_{12}L_{12}}\left(\frac{g_{m12}}{g_{m14}}\right)^2 + 2\frac{\rho_p}{W_{14}L_{14}}\right)\right.$$

$$\left. + 2\left(\gamma_n\left(\frac{g_{m12}}{g_{m14}^2}\right) + \frac{\gamma_p}{g_{m14}}\right)\right] \tag{5.50}$$

5.2.4   Power supply rejection ratio

5.2.4.1   Rejection of V_{DD} noise

The V_{DD} noise in the transconductor in Fig. 5.1 is coupled to the output through two different paths: on one hand, M10 and M9 sources and bulks as well as M8 and M7 bulks are connected to V_{DD}; on the other hand, also one of the inputs in transistors, M1, M2, M3, M4, is connected to V_{DD} aiming to reduce the effective threshold voltage. Figure 5.10 shows a low/mid frequencies equivalent circuit for PSRR+ calculations.

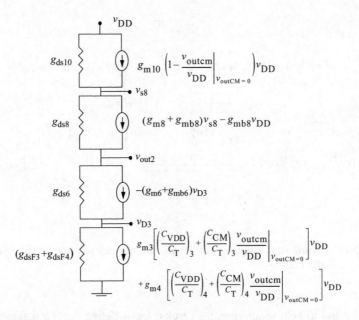

Figure 5.10   Small signal equivalent circuit for the PSRR+

The voltage gain at the output is given by eq. (5.51)[47]:

$$
\left.\frac{v_{out}}{v_{DD}}\right|_{v_{ind}=0}
$$

$$
\approx -\frac{[g_{m3}(C_{CM}/C_T)_3 + g_{m4}(C_{CM}/C_T)_4 - g_{m1}(C_{CM}/C_T)_1 - g_{m2}(C_{CM}/C_T)_2]}{G_{outn} + G_{outp}}
$$

$$
\times \left(\left.\frac{v_{outcm}}{v_{DD}}\right|_{v_{outCM}=0}\right)
$$

$$
- \frac{[g_{m3}(C_{VDD}/C_T)_3 + g_{m4}(C_{VDD}/C_T)_4 - g_{m1}(C_{VDD}/C_T)_1 - g_{m2}(C_{VDD}/C_T)_2]}{G_{outn} + G_{outp}}
$$

$$
+ \frac{(g_{m10} - g_{m9})[1 - v_{outcm}/v_{DD}|_{v_{outCM}=0}]}{G_{outn} + G_{outp}} \tag{5.51}
$$

A first conclusion that can be extracted from the last term in eq. (5.51) is that if the gain $(v_{outcm}/v_{DD})|_{v_{outCM}=0}$ is positive (as is the case) then the common mode feedback at the gate of the top p-channel transistors is going to contribute to reduce this term and therefore to increase the PSRR+. On the contrary, the feedback at the bottom n-channel FGMOS transistors will have the opposite effect. For the sake of

---

[47] The term corresponding to M8 bulk will be much smaller.

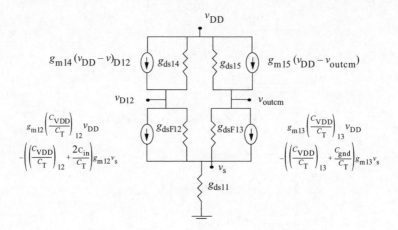

Figure 5.11   Small signal equivalent circuit for the calculation of the PSRR+ in the CMFB

simplicity and to help to interpret the results, let us define again a parameter Δ[48] to account for the mismatch and let us consider as a first approach only mismatch between the capacitances:

$$\left(\frac{C_{VDD}}{C_T}\right)_1 = \frac{C_{VDD}}{C_T} \qquad \left(\frac{C_{VDD}}{C_T}\right)_{i=[2,4]} = \frac{C_{VDD}}{C_T}(1 + \Delta_{3i}) \qquad (5.52)$$

Then, the final expression for eq. (5.51) is[49]

$$\frac{v_{out}}{v_{DD}}\bigg|_{v_{ind}=0} \approx \left(-g_{m3}\left(\frac{C_{CM}}{C_T}\right)[\Delta_{23} - \Delta_{22} + k_1\Delta_{24}]\left(\frac{v_{outcm}}{v_{DD}}\bigg|_{v_{outCM}=0}\right)\right.$$

$$\left. - g_{m3}\left(\frac{C_{VDD}}{C_T}\right)[\Delta_{33} - \Delta_{32} + k_1\Delta_{34}]\right)(G_{outn} + G_{outp})^{-1}$$

$$(5.53)$$

The contribution of the CMFB (Fig. 5.11) under the assumptions of no body effect in the input transistors, only mismatch between the input capacitances ($g_{m13} = g_{m12}$), a much smaller parasitic capacitance between the FG and source than the total

---

[48] As in the CMRR section.
[49] If a similar percentage of mismatch is considered for the transconductances as well, the final result for eq. (5.53) would approximately double.

capacitance seen by the FG and $g_{dsF} \ll g_m$, is

$$\left. \frac{v_{outcm}}{v_{DD}} \right|_{v_{outCM}=0} \approx 1 - \left(\left[\left(\left(\frac{C_{VDD}}{C_T} \right)_{13} \left(\left(\frac{C_{VDD}}{C_T} \right)_{12} + \frac{2C_{in}}{C_T} \right) \right. \right. \right.$$

$$\left. \left. - \left(\frac{C_{VDD}}{C_T} \right)_{12} \left(\left(\frac{C_{VDD}}{C_T} \right)_{13} + \frac{C_{gnd}}{C_T} \right) \right] g_{m13} \right)$$

$$\left. \times \left(g_{m15} \left[\left(\frac{C_{VDD}}{C_T} \right)_{12} + \frac{2C_{in}}{C_T} + \left(\frac{C_{VDD}}{C_T} \right)_{13} + \frac{C_{gnd}}{C_T} \right] \right)^{-1} \right)$$

$$\tag{5.54}$$

If the power supplies have been used to generate the reference voltage, the second term in eq. (5.54) will always be positive which reduces the gain. Otherwise it will be zero, unless there is mismatch. The final PSRR+ for the transconductor block is given by

$$\text{PSRR+} \approx (k_1 - 1) \left(- \left(\frac{C_{CM}}{C_{in}} \right) [\Delta_{23} - \Delta_{22} - k_1 \Delta_{24}] \left(\left. \frac{v_{outcm}}{v_{DD}} \right|_{v_{outCM}=0} \right) \right.$$

$$\left. - \left(\frac{C_{VDD}}{C_{in}} \right) [\Delta_{33} - \Delta_{32} - k_1 \Delta_{34}] \right)^{-1} \tag{5.55}$$

Equation (5.55) shows how the PSRR+ improves as the contribution of the input capacitance to the total capacitance increases and the values of the capacitances connected to the CMFB and V_{DD} decrease.

Let us now analyse as an example the limit case $k_1 \rightarrow 0$, or what is the same $G_m = (-C_{in}/C_T)g_{m3}$. The equation for the PSRR+ in this case is

$$\text{PSRR+} \approx \frac{1}{(C_{CM}/C_{in})[\Delta_{23} - \Delta_{22}](v_{outcm}/v_{DD}|_{v_{outCM}=0}) + (C_{VDD}/C_{in})[\Delta_{33} - \Delta_{32}]}$$

$$\tag{5.56}$$

Hence, for example, in the hypothetical case of a design with a maximum of 1 per cent mismatch between capacitances and $C_{CM} = C_{in} = 0.2C_T$, $C_{VDD} = 0.4C_T$, the worst PSRR+ predicted by eq. (5.56) would be 25 dB.

Summarising the main sources of degradation of the PSRR+ are: (a) The V_{DD} connection at the source of the load transistors in the CMFB, which is coupled to the main circuit block through the capacitance C_{CM} in the input FGMOS devices. (b) The connection to V_{DD} used to bias the FGMOS input devices in the right operating region. Hence, even when by doing this the topology is greatly simplified and the power reduced the price to pay in terms of PSRR+ should be kept in mind when establishing the design trade-offs.

5.2.4.2   Rejection of V_{SS} noise

The rejection of the negative power supply noise can be estimated with a similar circuit to the one in Fig. 5.10 but with V_{SS} instead of V_{DD} as the input, and also a different input coming from the CMFB ($v_{outcm}/v_{SS}|_{voutCM=0}$):

$$\text{PSRR}- \approx \left(\left(\frac{C_{in}}{C_T} \right) (g_{m4} - g_{m3}) \right)$$

$$\times \left(\sum_{i=3}^{4} \left[1 - \left(\frac{C_{CM}}{C_T} \right)_i \left(\frac{v_{outcm}}{v_{SS}} \Big|_{voutCM=0} \right) \right] g_{mi} \right.$$

$$\left. - \sum_{i=1}^{2} \left[1 - \left(\frac{C_{CM}}{C_T} \right)_i \left(\frac{v_{outcm}}{v_{SS}} \Big|_{voutCM=0} \right) \right] g_{mi} \right)^{-1} \qquad (5.57)$$

Considering now, as before, only mismatch between capacitances:

$$\text{PSRR}- \approx -\frac{(k_1 - 1)}{(C_{CM}/C_{in})[\Delta_{23} - \Delta_{22} + k_1 \Delta_{24}]((v_{outcm}/v_{SS})|_{voutCM=0})} \qquad (5.58)$$

where the value of $(v_{outcm}/v_{SS})|_{voutCM=0}$, is the same as for a normal differential pair with MOS transistors, unless that the negative power supply (V_{SS}) is also used to generate the reference in the CMFB in which case it would be

$$\frac{v_{outcm}}{v_{SS}} \Big|_{voutCM=0} \approx -\left(\frac{C_{gnd}}{2C_T} \right) \frac{g_{m12}}{g_{m15}} \qquad (5.59)$$

Hence, for example, for the same limit condition as before, $G_m = (-C_{in}/C_T)g_{m3}$, and the same hypothetical assumptions together with $C_{gnd} = 0.25\,C_T$, the PSRR− for a 1 per cent maximum mismatch would be 52 dB. Equation (5.58) predicts a degradation of the PSRR− as the absolute value of the transconductance decreases (due to the lower differential gain).

Summarising, using n-channel FGMOS devices is specially critical for the PSRR+, if the positive power supply voltage is used to bias them in the strong inversion saturation region. However, if p-channel input devices biased by the negative power supply voltage are employed instead, the PSRR+ would not be so critical but the PSRR− would be worse.

5.2.5 A filter example

This section illustrates the design of a lowpass/bandpass biquad filter that uses the integrator in Fig. 5.1 as the main building block. The filter is described in the time domain by the following state-space equations:

$$\dot{x}_1 = -\omega_{o2}x_1 - \omega_{o1}x_2 + \omega_{o4}u \qquad (5.60)$$

$$\dot{x}_2 = \omega_{o3}x_1 \qquad (5.61)$$

where x_1, x_2 and u are the bandpass output, lowpass output and input, respectively. Four coefficients, ω_{o1}, ω_{o2}, ω_{o3} and ω_{o4}, define the filter programmability. The cutoff frequency ω_o, quality factor Q, lowpass DC gain $H_{LP}(0)$ and bandpass gain at the

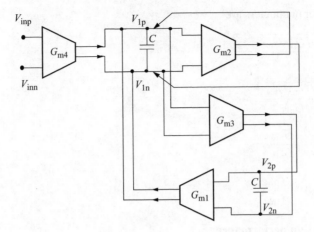

Figure 5.12   Block diagram for a second order G_m–C filter circuit

cut-off frequency, $H_{BP}(\omega_o)$, depend on these coefficients as follows:

$$\omega_o = \sqrt{\omega_{o3}\omega_{o1}} \tag{5.62}$$

$$Q = \frac{\sqrt{\omega_{o1}\omega_{o3}}}{\omega_{o2}} \tag{5.63}$$

$$H_{LP}(0) = \frac{\omega_{o4}}{\omega_{o1}} \tag{5.64}$$

$$H_{BP}(\omega_o) = \frac{\omega_{o4}}{\omega_{o2}} \tag{5.65}$$

A differential form of state space equations (5.60) and (5.61) is given by

$$\dot{x}_{1p} - \dot{x}_{1n} = -\omega_{o2}(x_{1p} - x_{1n}) - \omega_{o1}(x_{2p} - x_{2n}) + \omega_{o4}(u_p - u_n) \tag{5.66}$$

$$\dot{x}_{2p} - \dot{x}_{2n} = \omega_{o3}(x_{1p} - x_{1n}) \tag{5.67}$$

The circuit that implements these equations is shown in Fig. 5.12, where

$$\frac{x_{1p} - x_{1n}}{u_p - u_n} = \frac{V_{1p} - V_{1n}}{V_{inp} - V_{inn}} = \frac{V_{BP}(s)}{V_{ind}(s)} = \frac{s(G_{m4}/C)}{s^2 + sG_{m2}/C + (G_{m3}G_{m1})/C^2} \tag{5.68}$$

$$\frac{x_{2p} - x_{2n}}{u_p - u_n} = \frac{V_{2p} - V_{2n}}{V_{inp} - V_{inn}} = \frac{V_{LP}(s)}{V_{ind}(s)} = \frac{(G_{m3}G_{m4})/C^2}{s^2 + sG_{m2}/C + (G_{m3}G_{m1})/C^2} \tag{5.69}$$

And the filter parameters are

$$\omega_0 = \sqrt{\frac{G_{m3}G_{m1}}{C^2}} \tag{5.70}$$

$$Q = \frac{\sqrt{G_{m3}G_{m1}}}{G_{m2}} \tag{5.71}$$

$$H_{LP}(0) = \frac{G_{m4}}{G_{m1}} \tag{5.72}$$

$$H_{BP}(\omega_0) = \frac{G_{m4}}{G_{m2}} \tag{5.73}$$

5.2.5.1   Design performance

The integrator and filter are designed in a 0.8 μm double poly CMOS technology with threshold voltages in the order of 0.8 V [214]. The design is intended for audio applications and again the main goals are to reduce the values of power consumption and supply voltage. The time constants $(G_m/2\pi C)$ required for a frequency range between 100 Hz to 10 kHz are in the interval [100 μs, 10 ms]. As the time constant is inversely proportional to the integrating capacitance and proportional to the transconductance, its lowest required value will determine either the minimum transconductance or the maximum capacitance values that will be needed in the design. Sometimes, it is very difficult to realise very low G_ms values which forces the designer to choose a high capacitance value instead, hence compromising the area. If the area is a constraint, the design will have to be programmable to achieve the low $_m$ values, which is much more challenging. For our design, a relatively low capacitance value of 5 pF is chosen. This means the G_m needs to be programmable in the range [3 ns, 0.3 μs].

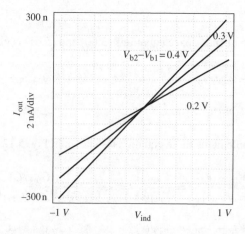

Figure 5.13 G_m-tuning for $(V_{b2} - V_{b1}) \sim 0.2, 0.3$ and 0.4 V

Transistors M1–M4 are designed with an aspect ratio $W/L = 2.5\ \mu m/10\ \mu m$ and input capacitances of 150 fF and 250 fF (for C_{VDD}). M12 and M13 aspect ratio is 2.5 $\mu m/17\ \mu m$ and their input and total capacitances are 133 fF and 533 fF, respectively. Figure 5.13 and Fig. 5.14 show simulation results for the output current (I_{out}) versus the differential input voltage for several values of the programming voltage ($V_{b2} - V_{b1}$). As predicted by eq. (5.3) the transconductor is very linear in the whole operating range. The exact characterisation in terms of linearity is given by the THD test. The results of this test for $G_{m1} = 3$ ns and $G_{m2} = 300$ ns are shown in Fig. 5.15(a) and (b), respectively. Both have been obtained at the edges of the input common mode

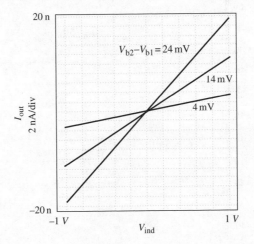

Figure 5.14 G_m-tuning for $(V_{b2} - V_{b1}) \sim$ 4, 14 and 24 mV

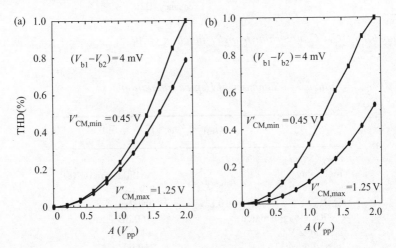

Figure 5.15   THD versus peak-to-peak input amplitude: (a) for $G_m = 3$ ns, and (b) for $G_m = 300$ ns

range (the difference between both common mode voltages is 0.8 V). The total THD is always below 1 per cent, and again the results are in agreement with the theoretical ones in eq. (5.45). The lossless integrator transfer functions for two transconductance values is shown in Fig. 5.16. In both cases, the integrator losses are below 20 Hz, far enough from the audio frequency range. Table 5.3 summarises the performance of the integrator for these two transconductance values and Fig. 5.17 illustrates different bandpass and lowpass filter transfer functions within the programmability range.

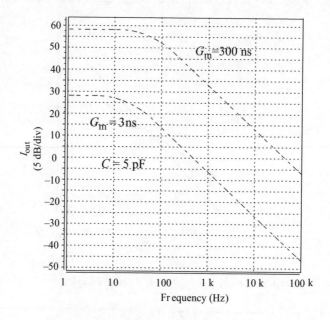

Figure 5.16 G_m-C integrator transfer-function [120]

Table 5.3   Summary of Hspice simulations

	$G_{m1} = 3$ ns	$G_{m2} = 300$ ns
V_{DD}		2 V
Technology		AMS0.8 μm (CXQ)
V_{ind}	2 V_{pp}	2 V_{pp}
CMR	0.8 V	0.8 V
CMRR (1% mismatch)		>60 dB
THD @ 2 V_{pp}		<1%
Inband noise		210 μV
Power		10 μW

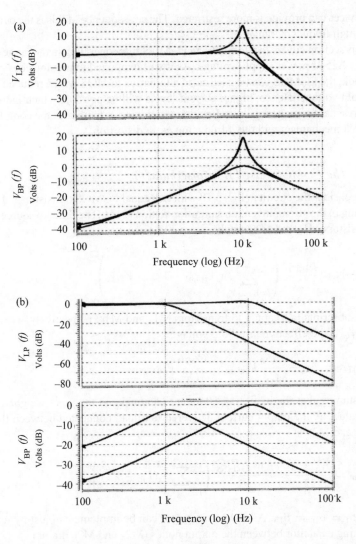

Figure 5.17   Programming of the filter. (a) Q-programming. (b) f_0-programming

5.3   An intermediate frequency FGMOS-based filter

Monolithic filters in the frequency range of 100 kHz to 10 MHz have many applications in communication receivers and video processing as well as in data communications and local area networks [147,150–155,157–159,166]. In general, design techniques that are applicable to voice band applications do not perform properly in this frequency range. They become even more difficult to extrapolate as the voltage supply limits are pushed down, because the maximum values of the currents that devices can provide is more restricted. The consequence of this is the degradation of the high frequency range.

This section covers the design of an FGMOS filter operating in the range of megahertz at 1.25 V power supply voltage. As it was explained in previous designs along this book, the FGMOS can be used as an alternative to MOS with a reduced effective threshold voltage. Hence, higher current levels will be possible which increases the upper reachable frequency limit. Besides, the transistor facilitates the common mode feedback mechanism without introducing extra distortion.

5.3.1   The FGMOS-based IF transconductor

Let us consider the circuit in Fig. 5.18 with two equal FGMOS transistors ($M1_a$, $M1_b$) operating in the strong inversion saturation region [215]. The drain to source currents of transistors $M1_a$ and $M1_b$ can be expressed as[50]

$$I_{1(a,b)} = \frac{\beta_{1(a,b)}}{2} \left(\sum_{i=1,\ldots,N} w_i V_{i(a,b)} - V_{S1} - V_{TFG} \right)^2 \tag{5.74}$$

where subscripts (a,b) apply to transistors $M1_a$ and $M1_b$, respectively, $w_i = C_i/C_T$, and V_{TFG} is the effective threshold voltage:

$$V_{TFG} = V_{Tn} - \frac{C_{CM}}{C_T} V_{outcm} - \frac{C_{VDD}}{C_T} V_{DD} \tag{5.75}$$

Assuming that the input transistors are identical and the input weights have the same value, w, the output current delivered to a load connected between the drains of $M2_a$ and $M2_b$ is

$$I_{out} = \frac{I_{1b} - I_{1a}}{2} \approx \frac{w\sqrt{\beta_1 I_{SS}}}{2} \left(\sum_{i=1,\ldots,N} (V_{ib} - V_{ia}) \right) \tag{5.76}$$

where $\beta_1 = \beta_{1a} = \beta_{1b}$. A lossless integrator can be implemented just by joining an integrating capacitor between the output nodes ($M2_a$ and $M2_b$ drains).

This approximation has the same limitations as a conventional differential pair [160]. The difference is that in this case the voltage that the FGs see is $w \cdot \sum_{i=1\ldots,N} V_{i1(a,b)}$ as opposed to $\sum_{i=1\ldots,N} V_{i(a,b)}$. This means that for a certain value of distortion the input range increases by a factor $1/w$. Equation (5.76) shows a linear relationship between voltages and current. The gain factor is the transconductance:

$$G_m = \frac{w\sqrt{\beta_1 I_{SS}}}{2} \tag{5.77}$$

---

[50] This equation is valid whenever the parasitic capacitances are negligible compared with the FGMOS inputs.

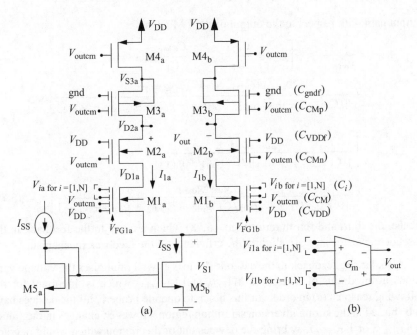

Figure 5.18 (a) Proposed FGMOS transconductor. (b) Symbol

5.3.2 The FGMOS cascode load and the CMFB

As Fig. 5.18 shows, the transconductor load consists of a cascode structure. The reasons that justify this, in spite of the fact that stacking transistors increases the value of the minimum voltage needed for the circuit to work, are:

1. The output resistance is improved and hence the losses reduced, which is even more important now than when using normal MOS devices, since the FGMOS conductance is higher due to the added $(C_{GD}/C_T)g_m$ term (see Chapter 2).

2. The cascode bias voltages can be generated within the same devices just by choosing the appropriate weights. The latter is equivalent to shifting the effective threshold voltages appropriately so that the transistors can be biased using already available voltages within the system. This avoids having to have extra circuitry for this aim (and hence, extra power) or very large transistors with also large parasitic capacitances that would affect the frequency response.

3. As it was already explained before, because the topology is differential a CMFB is required [216]. The cascode devices can help to improve the efficiency of the feedback mechanism in the following way: if V_{outcm} is the output voltage in the CMFB and this is fed back to the cascode transistors through the input capacitances C_{CMn} (in the n-channel cascode device) and C_{CMp} (in the p-channel cascode device) as well as to the input and load transistors (M1$_{(a,b)}$ and M4$_{(a,b)}$, respectively) the

output gain with respect to the output of the CMFB is[51]

$$A_{CM} = \frac{v_{D2}}{v_{outcm}} \approx -\left[g_{m4} + \frac{C_{CM}}{C_T}g_{m1} + \frac{C_{CMp}}{C_T}g_{ds4} \right.$$

$$\left. + \frac{(C_{CMn}/C_T) \cdot g_{m2}g_{dsF1}}{[(C_{CMn}/C_T + C_{VDDf}/C_T)g_{m2} + g_{mbF2} + g_{dsF2} + g_{dsF1}]} \right] R_{out}$$

(5.78)

$$R_{out} \approx \left[\frac{g_{dsF3}g_{ds4}}{[(C_{CMp}/C_T + C_{gndf}/C_T)g_{m3} + g_{dsF3}]} \right.$$

$$\left. + \frac{g_{dsF2}g_{dsF1}}{[(C_{CMn}/C_T + C_{VDDf}/C_T)g_{m2} + g_{mbF2} + g_{dsF2} + g_{dsF1}]} \right]^{-1}$$

(5.79)

Hence, the third and fourth term in eq. (5.78) which are due to the feedback in the cascode transistor increase slightly the efficiency of the feedback mechanism.

Also, having feedback in the cascode devices as well minimises the voltage variations in the drains of transistors $M1_{(a,b)}$ and $M4_{(a,b)}$, which is important for the following reasons: (a) in order for this block to operate properly all the devices have to be biased in the strong inversion saturation region. However, changes in the values of $V_{D1(a,b)}$ or $V_{S3(a,b)}$ may bring the devices out of this region which would degrade the performance of the whole block. (b) Also, from the point of view of $V_{D1(a,b)}$, keeping this voltage constant can help to minimise the distortion caused by the term corresponding to the gate to drain coupling in the transistor current law[52].

Analytically, the variations that $V_{D1(a,b)}$ and $V_{S3(a,b)}$ experience when the common mode feedback voltage changes are

$$\frac{\partial V_{S3(a,b)}}{\partial V_{outcm}} = \left(\frac{C_{CMp}}{C_T} - \sqrt{\frac{\beta_4}{\beta_3}} \right)$$

(5.80)

$$\frac{\partial V_{D1(a,b)}}{\partial V_{outcm}} = \left(\frac{C_{CMn}}{C_T} - \frac{C_{CM}}{C_T}\sqrt{\frac{\beta_1}{\beta_2}} \right)$$

(5.81)

where the first positive term in the equations (which compensates for the second negative term) is due to the aforementioned feedback in the cascode transistors.

The price to be paid for the previous advantages is related to the degradation of the PSRR because of the use of the power supply voltages to bias the cascode devices. This will be analysed in more detail in a different section.

The CMFB suggested for this block is the same as in the previous design. However, for the design example at the end of this section p-channel MOS transistors were used

---

[51] V_{D2a} stands for the single side output as appears in Fig. 5.18. Also, the subscripts (a,b) have been eliminated for now in the parameters names for the sake of simplicity as ideally parameters of the same type that only differ in the a, b subscripts are supposed to be identical. As usual subscript i alludes to device Mi.

[52] This term was neglected in eq. (5.74) for the sake of simplicity.

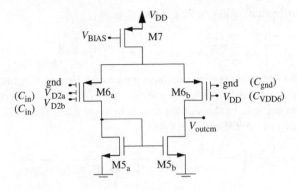

Figure 5.19   The CMFB

at the input[53]. The schematic with the values of the corresponding input capacitances is shown in Fig. 5.19.

5.3.2.1   The Common Mode Rejection Ratio (CMRR)

As it was previously explained for other transconductors blocks, the fully differential characteristic of the topology should reject all common mode signals. In reality mismatch will make the common mode gain different from zero though. In the case of this particular transconductor if only one effective input ($V_{in(a,b)}$) is assumed[54], the DC common mode gain under the assumption that matching is not perfect is

$$\left.\frac{v_{D2b} - v_{D2a}}{v_{inCM}}\right|_{v_{ind}=0} = \left.\frac{v_{out}}{v_{inCM}}\right|_{v_{ind}=0}$$

$$\approx \frac{-g_{m1b}(C_{in}/C_T)_b R_{outb}}{1 + |A_{CMb}|(2(C_{in}/C_T)g_{m6})/(g_{dsF6} + g_{ds5})}$$

$$+ \frac{g_{m1a}(C_{in}/C_T)_a R_{outa}}{1 + |A_{CMa}|(2(C_{in}/C_T)g_{m6})/(g_{dsF6} + g_{ds5})} \tag{5.82}$$

Subscripts a and b have been added to distinguish between the now different parameters in both branches (because of mismatch). It has also been assumed that the effective input weights in the CMFB are the same as in the main transconductor in absence of mismatch (C_{in}/C_T). Also, mismatch in the CMFB has been neglected, since it will not change significantly the final results but would make the equations however much less intuitive.

---

[53] It was more convenient in this case because of the characteristics of the technology used. It was easier this way to obtain adequate voltage levels.

[54] For the sake of clarity in the expressions. If the transconductor has more than one input the respective contributions add up. The equivalent input weight is C_{in}/C_T.

The DC differential gain can be expressed as

$$A_d = \frac{v_{D2b} - v_{D2a}}{(v_{inb} - v_{ina})} = \frac{v_{out}}{v_{ind}}\bigg|_{v_{inCM}=0} \approx -\frac{C_{in}}{C_T} \cdot g_{m1} \cdot R_{out} \tag{5.83}$$

Equation (5.83) does not take into account mismatch. This approximation has been made for the sake of clarity. Also, it is justified because the main effect of mismatch in the CMRR is going to be through the common mode gain. The CMRR is given by the ratio between eq. (5.83) and eq. (5.82):

$$\text{CMRR} \approx \left(\frac{C_{in}}{C_T}\right)\left(\frac{(C_{in}/C_T)_b}{1 + |A_{CMb}|[2(C_{in}/C_T)g_{m6}]/[g_{dsF6} + g_{ds5}]}\right.$$
$$\left. - \frac{(C_{in}/C_T)_a}{1 + |A_{CMa}|[2(C_{in}/C_T)g_{m6}]/[g_{dsF6} + g_{ds5}]}\right)^{-1} \tag{5.84}$$

The value predicted by eq. (5.84) can be as high as 80 dB if the relationship between the transconductances and conductances is larger than 10, the input weight is 0.5 and the mismatch is below 10 per cent.

5.3.3 Dynamic range

The transconductor in Fig. 5.18 can be seen as a modified version of a conventional differential pair. From the point of view of linearity the behaviour referred to the FGs ($FG1_a$ and $FG1_b$) is pretty much the same as in the normal differential pair. This is because of the cascode devices that keep $M1_a$ and $M1_b$ drain voltages almost constant, and hence reduce the nonlinear effect of the term $(C_{GD}V_{D1(a,b)})/C_T$ in the current equations. Therefore, the input range referred to the FGs would be the same. However, referred to the effective input, the input range for a certain level of distortion is amplified by the inverse of the effective input weight, this is by C_T/C_{in}.

Regarding the noise the total RMS spectral density of noise referred to one of the inputs is given by[55]

$$V_{inRMS} \approx \left(\frac{C_T}{C_{in}}\right)\sqrt{S_1 + S_4\frac{g_{m4}^2}{g_{m1}^2}} \tag{5.85}$$

where S_1 and S_4 are the power spectral density of noise for transistors $M1_{(a,b)}$ and $M4_{(a,b)}$, which are the main noise contributors.

Hence, taking into account that the equivalent input range for a certain level of distortion increases by C_T/C_{in} but also the equivalent input noise does, the final dynamic range would remain the same as in a MOS realisation. The advantage is that it is possible to keep the dynamic range and simultaneously have a large input range, which can be very important for certain applications specially under the design constraints of low power and low voltage[56].

---

[55] Other noise contributions are negligible.

[56] There is, for example, no need of designing extra circuitry to attenuate incoming signals, which would consume extra power.

5.3.4 Rejection to supply noise

5.3.4.1 Rejection to V_{DD} noise (PSRR+)

In the transconductor in Fig. 5.18, the V_{DD} noise is coupled to the output through transistors $M1_{(a,b)}$ and $M2_{(a,b)}$ inputs, $M4_{(a,b)}$ source, as well as from the feedback from the CMFB. Ideally, as it was already said before, considering that this structure is fully differential, both branches should be equally affected, and hence the difference between the two of them at the output should be zero. However, mismatch increases this difference, which is then given by

$$
\left. \frac{v_{out}}{v_{DD}} \right|_{v_{ind}=0} \approx \left[-g_{m1b}\left(\frac{C_{VDD}}{C_T}\right)_b R_{outb} + g_{m1a}\left(\frac{C_{VDD}}{C_T}\right)_a R_{outa} \right.
$$

$$
\left. + g_{m4b}R_{outb} - g_{m4a}R_{outa} \right] + (A_{CMb} - A_{CMa})\left. \frac{v_{outcm}}{v_{DD}} \right|_{v_{outCM}=0}
$$

$$
- \frac{(C_{VDDf}/C_T)_b \cdot g_{m2b}g_{dsF1b} \cdot R_{outa}}{[((C_{CMn}/C_T)_b + (C_{VDDf}/C_T)_b)g_{m2b} + g_{mbF2b} + g_{dsF2b} + g_{dsF1b}]}
$$

$$
+ \frac{(C_{VDDf}/C_T)_a \cdot g_{m2a}g_{dsF1a} \cdot R_{outa}}{[((C_{CMn}/C_T)_a + (C_{VDDf}/C_T)_a)g_{m2a} + g_{mbF2a} + g_{dsF2a} + g_{dsF1a}]}
\tag{5.86}
$$

where $(v_{outcm}/v_{DD})|_{v_{outCM}=0}$ is the output to v_{DD} gain for the CMFB:

$$
\left. \frac{v_{outcm}}{v_{DD}} \right|_{v_{outCM}=0} \approx \frac{-g_{m6}}{g_{dsF6} + g_{ds5}} \frac{C_{VDD6}}{C_T}
\tag{5.87}
$$

In order to get a more clear understanding of the FGMOS devices contribution to the PSRR+, let us again assume that the main deviation from the ideal values occurs in the input capacitances. Using eq. (5.83) for the differential gain, the PSRR+ is given by

PSRR+

$$
\approx \left(\frac{C_{in}}{C_T}\right)\left(\left(\frac{C_{VDD}}{C_T}\right)_b - \left(\frac{C_{VDD}}{C_T}\right)_a + \left[\left(\frac{C_{CM}}{C_T}\right)_b - \left(\frac{C_{CM}}{C_T}\right)_a\right]\left. \frac{v_{outcm}}{v_{DD}} \right|_{v_{outCM}=0}\right)^{-1}
\tag{5.88}
$$

Equation (5.88) shows that the most significant contributions of the FGMOS devices to the PSRR+ come from the input transistors in both the main transconductor block and the CMFB. In particular, if the gain of the CMFB is too high its corresponding PSRR+ term will become dominant and the value in eq. (5.88) will be seriously degraded. Again, a possible way to avoid this if it becomes a problem would be to generate the reference voltage in the CMFB separately, instead of directly using V_{DD} for it. As an example of the orders of magnitude for the PSRR+ in this structure, if the worst case of mismatch is 1 per cent, the CMFB gain is 10, $C_{VDD} = C_{in}$, $C_{VDD6} = 0.5C_T$ and $C_{CM} = 0.5C_{in}$ the PSRR would just be 29 dB. However, if V_{DD} is not used to generate the reference voltage in the CMFB, the PSRR+ would increase to 40 dB. Besides, these calculations have been carried out neglecting the mismatch between transconductances, which is not strictly right. Hence, for example, variations with respect to their ideal values of $M4_{(a,b)}$ transconductances, would generate a term of around the same value as the denominator in eq. (5.88). This would degrade the PSRR+ even more. The conclusion to this analysis is that the designer should be very cautious when designing a circuit like this if V_{DD} is used within a high gain CMFB to generate the voltage reference, if the PSRR+ is an issue. Whenever it is, alternative techniques should be used.

5.3.4.2 PSRR−

A similar analysis can be carried out to derive the effect of the negative power supply voltage noise[57] on the transconductor performance. Again, as eq. (5.89) shows, the PSRR will be highly dependent on the CMFB:

$$\left.\frac{v_{out}}{v_{SS}}\right|_{v_{ind}=0} \approx (A_{CMb} - A_{CMa})\left.\frac{v_{outcm}}{v_{SS}}\right|_{v_{outCM}=0} \tag{5.89}$$

where $(v_{outcm}/v_{SS})|_{v_{outCM}=0}$ is the output to v_{SS} gain for the CMFB. Under the same assumptions as before

$$\left.\frac{v_{outcm}}{v_{SS}}\right|_{v_{outCM}=0} \approx \frac{g_{m6}}{g_{ds5} + g_{dsF6}}\left[\left(\frac{C_{gnd}}{C_T}\right)_a - \left(\frac{C_{gnd}}{C_T}\right)_b\right] \tag{5.90}$$

Considering only mismatch between capacitances and using the eq. (5.83) for the differential gain

$$PSRR- \approx \frac{(C_{in}/C_T)}{((C_{CM}/C_T)_b - (C_{CM}/C_T)_a)(v_{outcm}/v_{SS})|_{v_{outCM}=0}} \tag{5.91}$$

As expected, the PSRR− is mostly affected by the ground connected input in the FGMOS transistors of the CMFB. As an example, the resulting PSRR− using the same values for the parameters as for the PSRR+ as well as $C_{gnd} = C_{in} = 0.5\,C_T$ would be 66 dB. But this is assuming that both of the capacitances connected to ground in the CMFB have identical nominal values. However, if they are different, and for example $C_{gndb} > C_{gnda}$ because C_{gndb} is also used to generate the reference voltage, the PSRR− value will be much smaller. Let us assume that $C_{gnda} = (2/3)C_{gndb}$. In this case, the PSRR− would be reduced to just 36 dB. Comparing these two numbers, the conclusions and recommendations are the same as for the PSRR+: if the negative power supply voltage noise is a concern, the use of V_{SS} should be avoided for the generation of the reference voltage in the CMFB and alternative techniques should be employed. However, using V_{SS} to bias the transistors within this block in the strong inversion saturation region is not much of a problem being still possible to achieve a reasonably good value for the PSRR−[58].

---

[57] v_{SS}, which in this case is ground.
[58] Again, this is assuming only mismatch between the input capacitances in the FGMOS devices. The contribution of the other terms must be added to these results for an accurate estimation of the PSRR.

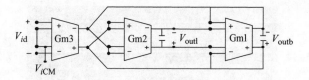

Figure 5.20   Second-order filter schematic

5.3.5   A second-order filter prototype

A second-order filter can be built using this transconductor as the core block as it is illustrated in this section. The chosen power supply voltage for our example is 1.25 V and the technology is a 0.35 μm CMOS process [217] with threshold voltages of $V_{Tn} = 0.45$ V and $V_{Tp} = -0.62$ V[59].

The starting point for the design is the state space equations:

$$\dot{x}_1 = -\omega_{o2}x_1 - \omega_{o1}x_2 + \omega_{o3}x_i \tag{5.92}$$

$$\dot{x}_2 = 2\omega_{o2}x_1 \tag{5.93}$$

Three ω_{oi} determine the filter programmability. The cut-off frequency, quality factor and gains depend on them as

$$\omega_o = \sqrt{2\omega_{o1}\omega_{o2}} \tag{5.94}$$

$$Q = \sqrt{2\frac{\omega_{o2}}{\omega_{o1}}} \tag{5.95}$$

$$H_{LP}(0) = \frac{\omega_{o3}}{\omega_{o1}} \tag{5.96}$$

$$H_{BP}(\omega_o) = \frac{\omega_{o3}}{\omega_{o1}} \tag{5.97}$$

The block diagram and symbol of the second-order filter is shown in Fig. 5.20, where

$$\omega_{oi} = \frac{G_{mi}}{C} = \frac{w}{2C}\sqrt{\beta_1 I_{SSi}} \tag{5.98}$$

As eq. (5.98) shows the ω_{oi} parameters can be modified by changing the tail current I_{SSi} in each transconductor. The signal V_{outb} is the output of the bandpass filter and V_{outl} the corresponding to the lowpass section.

The filter is designed using 1.5 pF integrating capacitors. The aspect ratio for the input transistors is 30 μm/0.35 μm. The input capacitances are 70 fF and the capacitance that controls the input threshold voltage shift is 140 fF. A microphotograph of the fabricated prototype is shown in Fig. 5.21. The total area is 0.05 mm². Examples of the circuit performance are shown from Fig. 5.22 to Fig. 5.24, but a more detailed

---

[59] Similar results were obtained in a 0.8 μm process [214] with threshold voltages $V_{Tn} = 0.82$ V and $V_{Tp} = -0.8$ V, with the exception of the area and the tuning range that moves towards lower frequencies.

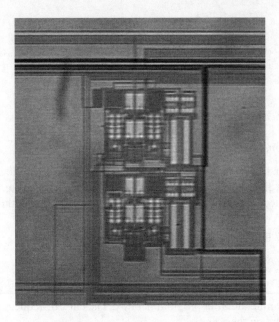

Figure 5.21   Microphotograph of the filter prototype

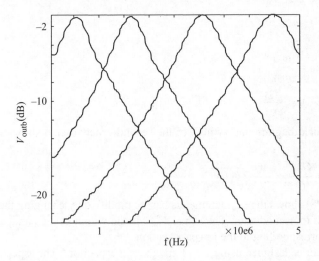

Figure 5.22   Cut-off frequency tuning ($I_{ss}(\mu A) = [4.7, 100]$)

description can be found in Table 5.4. The filter can be programmed in 0.7 decades being the maximum power consumption 0.5 mW for the maximum frequency. The DR is larger than 40 dB, being the maximum signal, for a THD of 1 per cent, 1.2 V_{pp}. In any case, this is not the optimum design but just an example of typical results that can be achieved with this kind of topology.

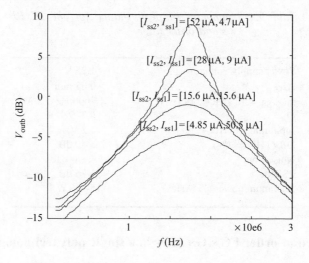

Figure 5.23   Quality factor tuning[60]

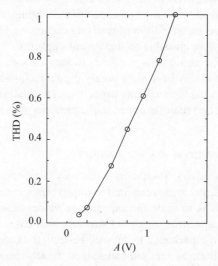

Figure 5.24   THD versus 100 kHz differential input for the lowpass filter with 1 MHz cut-off frequency and quality factor of 1

The results in Table 5.4 demonstrate the suitability of the FGMOS technique for intermediate frequencies. Again, by combining the extra degrees of freedom that the FGMOS transistor offers it is possible to push down the power supply voltage in circuits for IF frequencies applications while keeping a good DR.

---

[60] I_{ss2} and I_{ss1} are the bias currents used to tune ω_{o2} and ω_{o1}, respectively.

Table 5.4   Summary of performance for the IF BP second-order filter

Power supply voltage	1.25 V
Area	0.05 mm^2
$f_o@I_{SS} = [2.75\,\mu A,\ 145\,\mu A]$	700 kHz–5.2 MHz
Q	0.7–5
Input range (THD < 1%)	1.2 V_{pp}
DR (THD = 1%)	>40 dB
Noise floor	<−60 dB
PSRR	>36 dB
Maximum power@5 MHz	543 μW

5.4   A second-order FGMOS filter in a single poly technology

Until now, the only technological requirement for the design of circuits based on FGMOS transistors has been to have a double poly technology. This is because these circuits' functionality is based on capacitive couplings and using double poly is the easiest and more accurate form of realising capacitors. However, theoretically speaking it could also be possible to implement capacitors using metal on top of polysilicon (poly). The aim of this section is to analyse some of the consequences and somehow the feasibility of using a purely digital technology, without a second layer of poly and low quality metal/poly capacitors. The analysis uses as example the transconductor block and filter described in the previous section.

5.4.1   Effects of the metal–poly capacitors

The main drawback of having low quality metal/poly capacitors in FGMOS devices is the fact that a much larger area is required to implement a certain capacitance value. An alternative to this is to scale the capacitance values down while still keeping the effective inputs capacitive weights. However, both cases suffer from the same problem: the parasitic capacitance to the well beneath the capacitor might be of the same order of magnitude as the nominal values, which causes a reduction of the transistor effective transconductance. Let us express the effective input weight as

$$\frac{C_i}{C_T} \approx \frac{k_i A_i C_{pma} + k_i P_i C_{pmp}}{C_{Ta} + C_{Tp}} \tag{5.99}$$

$$C_{Ta} \approx (1 + k_i)A_i(C_{pma} + C_{pwa}) + k_{ca}C_{ox}A_t \tag{5.100}$$

$$C_{Tp} \approx (1 + k_i)P_i(C_{pmp} + C_{pmp}) + k_{cp}C_{ox}P_t \tag{5.101}$$

$$C_{ox} = \frac{\varepsilon_{ox}}{t_{ox}} \tag{5.102}$$

Table 5.5   *Typical capacitance-related*
technological parameters

C_{pma} (fF/μm^2)	0.054
C_{pmp} (fF/μm)	0.05
C_{pwa} (fF/μm^2)	0.12
C_{pwp} (fF/μm)	0.05

where C_{pma} is the poly1 to metal capacitance per unit of area, C_{pmp} per unit of perimeter, and C_{pwa} and C_{pwp} are the poly1 to well capacitance per unit of area and perimeter, respectively. Typical values of these parameters for a 0.35 μm technology are shown in Table 5.5. A_t and P_t are the area and perimeter of the transistor, and $(k_i A_i)$ and $(k_i P_i)$ are the area and perimeter of the equivalent input capacitances. A_i is the area of the capacitance used to get a certain effective threshold voltage. Parameters k_{ca} and k_{cp} depend upon the transistor operating region, and are related to the intrinsic parasitic capacitances. The other parameters have their usual meanings. The values in Table 5.5 show how in, for example this particular technology, the parasitic to the well will affect, at least, as a reduction of around a factor of two in the transconductance value.

A problem added to this is that the variations with respect to their nominal values in metal/poly capacitances can also be much bigger than in poly2/poly1 capacitances. An example of typical deviation values is 60 per cent for metal/poly1 capacitances versus 10 per cent for poly2/poly1.

The following sections analyse some of the most important consequences of reducing the values of the input capacitances in the transconductor in Fig. 5.18.

5.4.1.1   Offset

One of the main sources of systematic offset in the transconductor in Fig. 5.18 is the capacitive coupling between gate and drain in the input transistors. The voltage across the drain of transistors M1$_{(a,b)}$, M2$_{(a,b)}$ obtained by using the fact that their currents are the same is

$$V_{D1} \approx \left(V_{cmN} + \frac{C_{GD2}}{C_{T2}} V_{D2} + \sqrt{\frac{\beta_1}{\beta_2}} V_{Tn1} - V_{Tn2} \right)$$

$$\times \left(\frac{1}{1 - C_{GS2}/C_{T2} + (C_{GD1}/C_{T1})\sqrt{\beta_1/\beta_2}} \right) \tag{5.103}$$

This voltage will couple to the FG of the input transistor through C_{GD} giving rise to an equivalent input offset[61]:

$$V_{off+} \approx \frac{(C_{GD1}/C_{in})(V_{cmN} + \sqrt{\beta_1/\beta_2}V_{Tn1} - V_{Tn2})}{[A' - A_d(C_{GD2}/C_{T2}) \cdot (C_{GD1}/C_{in})]} \tag{5.104}$$

A_d being the transconductor output gain, and

$$V_{cmN} = \frac{C_{VDDf}}{C_{T2}}V_{DD} + \frac{C_{CMn}}{C_{T2}}V_{outcm} - \sqrt{\frac{\beta_1}{\beta_2}}\frac{C_{in}}{C_{T1}}V_{iCM} - \sqrt{\frac{\beta_1}{\beta_2}}\frac{C_{CM}}{C_{T1}}V_{outcm}$$

$$- \sqrt{\frac{\beta_1}{\beta_2}}\frac{C_{VDD}}{C_{T1}}V_{DD} + \sqrt{\frac{\beta_1}{\beta_2}}\left(1 - \frac{C_{GS1}}{C_{T1}}\right)V_{S1} \tag{5.105}$$

$$A' = \frac{1}{1 - C_{GS2}/C_{T2} + (C_{GD1}/C_{T1})\sqrt{\beta_1/\beta_2}} \tag{5.106}$$

Subscripts have been used to refer to the name of the transistor. In this case, since the absolute values of the capacitances can change a lot, different total capacitances have been considered. The differential offset can be obtained by subtracting two equations with the form of eq. (5.103) for both differential branches, a and b[62]:

$$\text{offset} \approx \left[\frac{(C_{GD1a}/C_{ina}) \cdot t_a}{(A'_a - A_{da}(C_{GD2a}/C_{T2})(C_{GD1a}/C_{ina}))} \right.$$

$$\left. - \frac{(C_{GD1b}/C_{inb}) \cdot t_b}{(A'_b - A_{db}(C_{GD2b}/C_{T2})(C_{GD1b}/C_{ina}))} \right] \tag{5.107}$$

where

$$t_{(a,b)} = \left(V_{cmN(a,b)} + \sqrt{\frac{\beta_{1a}}{\beta_{2a}}}V_{Tn1(a,b)} - V_{Tn2(a,b)}\right) \tag{5.108}$$

Ideally t_a and t_b should be equal but, in practice, any mismatch affecting any of the three terms in eq. (5.108) generates differences between the two of them. A new variable del1 will account for these variations as shown in eq. (5.109):

$$t_b = (1 + \text{del } l)t_a \tag{5.109}$$

Equation (5.107) shows how as C_{GD} becomes smaller in comparison to the input capacitance the offset decreases as well. The parameters that affect the offset the most are the common mode at the input, and the variations between parameters related to the term t_b, as for example threshold voltages or βs. But also, the mismatch between the input capacitances can be crucial. These conclusions are illustrated in Fig. 5.25. The nominal values used to obtain these graphs are the same as for the design example at the end of the section: $C_{GD} = 3.3$ fF, $C_{T1} = 81$ fF, $C_i = 11.15$ fF.

---

[61] A single effective input with associated capacitance C_{in} has been considered for the sake of simplicity.
[62] The subscripts a and b have been added to distinguish between both.

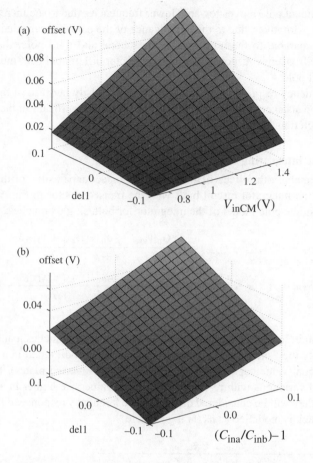

Figure 5.25 (a) *Offset versus common mode at the input and mismatch affecting threshold voltages and β. (b) Offset versus mismatch in the input capacitances and also versus mismatch affecting the threshold voltages and β*

5.4.1.2 Frequency response of the OTA

An exact analysis of the frequency response is complex and does not provide so much information, so this section just describes very briefly what are the differences between the frequency response of the transconductor built with metal/poly capacitors and the same transconductor with two layers of poly capacitors instead.

As before, the DC gain of the transconductor block is given by eq. (5.83). However, the output resistance given by eq. (5.79) is now much smaller, because of the increased output conductance of the FGMOS devices since the ratio C_{GD}/C_T is larger. Also, the effective transconductance is lower because of the large parasitic term added to the value of C_T. All this contributes to reduce the DC gain.

The dominant pole moves towards lower frequencies due to the increased output conductance. Moreover, due to the significance of the parasitic capacitances (which are now comparable to the input ones), the second and third pole, that normally appear in the frequency response of a transconductor differential pair, move closer to the dominant pole.

The frequency response of the CMFB changes mostly because of the reduction in its output resistance. The term $C_{GD6}g_{m6}/C_T$ is more significant in the expression of g_{dsF6} which translates to a reduction of the amplifier gain.

5.4.1.3   The integrating capacitance

This section analyses the effects of the integrating capacitor parasitics on the frequency response of the integrator circuit built with the transconductor in Fig. 5.18. Let us start with a single pole model of the integrator for both single branches:

$$sC_{pa}v_{D2a} + sC(v_{D2a} - v_{D2b}) = -\frac{G_m v_{id}}{2} - \frac{a_0}{s+p}\frac{(v_{D2a} + v_{D2b})}{2} - G_{outa}v_{D2a}$$

$$sC_{pb}v_{D2b} + sC(v_{D2b} - v_{D2a}) = \frac{G_m v_{id}}{2} - \frac{a_0}{s+p}\frac{(v_{D2a} + v_{D2b})}{2} - G_{outb}v_{D2b}$$

$$(5.110)$$

where C_{pa} and C_{pb} represent the parasitic capacitances in branches a and b, respectively. Ideally they should be the same if the capacitor is designed in a differential way (two equal value capacitors in parallel but with inverted plates), but because the expected variations with respect to their ideal values are higher in a metal/poly realisation, they will be considered different. The frequency response of the common mode feedback is modelled as $a_0/(s + p)$ where

$$p \approx \frac{g_{dsF6b} + g_{ds5b}}{\sum_{i=(a,b)}(C_{CMi} + C_{CMni} + C_{CMpi} + C_{GS4i})} \tag{5.111}$$

$$a_0 \approx \frac{C_{in}}{C_T} \times \frac{(g_{m1}(C_{CM}/C_T) + g_{m4})g_{m6}}{(C_{CM} + C_{CMn} + C_{CMp} + C_{GS4})G_{out}} \tag{5.112}$$

The common and differential gain can be obtained by solving the systems of equations (5.110). Assuming $G_{outa} = G_{outb} = G_{out}$

$$\frac{v_{outCM}}{v_{id}} = \frac{G_m(C_{pa} - C_{pb})(s+p)s}{2D(s)} \tag{5.113}$$

where

$$v_{outCM} = \frac{v_{d2a} + v_{d2b}}{2} \tag{5.114}$$

Equation (5.113) shows how the larger the mismatch between both parasitic capacitances, C_{pa} and C_{pb}, is the bigger the common mode gain at the output. In the worst case scenario, this could originate premature clipping.

The differential gain is given by

$$v_{out} = v_{d2b} - v_{d2a} \tag{5.115}$$

$$\frac{v_{out}}{v_{id}} = G_m \frac{(s+p)(2G_{out} + s(C_{pa} + C_{pb})) + 2a_o}{D(s)} \tag{5.116}$$

$$D(s) = 4a_o s \left(C + \frac{C_{pa} + C_{pb}}{4} + \frac{G_{out}^2}{2a_o} \right) + 2s^2(s+p) \left(C_{pa} + C_{pb} + \frac{C_{pa} C_{pb}}{C} \right) C$$
$$+ 2[G_{out}(C_{pa} + C_{pb}) + 2G_{out}C](s+p)s + 2G_{out}(G_{out}p + a_o) \tag{5.117}$$

Equation (5.116) shows two parasitic zeros in the integrator located at

$$z_{1,2} = -\frac{p}{2} - \frac{G_{out}}{(C_{pa} + C_{pb})} \pm \frac{1}{2} \sqrt{p^2 + \frac{4G_{out}^2}{(C_{pa} + C_{pb})} - 4\frac{(2a_o + G_{out}p)}{(C_{pa} + C_{pb})}} \tag{5.118}$$

Also, the dominant pole shifts towards a different frequency:

$$w_{p1}' \approx \frac{G_{out}(G_{out}p + a_o)}{2a_o(C + (C_{pa} + C_{pb})/4)} \tag{5.119}$$

The previous analysis applies to the integrator irrespectively of whether the integrating capacitance is implemented in either poly1/poly2 or metal/poly. The difference is that in the metal/poly implementation a_o is expected to be smaller because of the larger contribution of the parasitics to the value of C_T. Besides, the integrator will have extra losses (as predicted in eq. (5.119)) that might affect in more or less degree[63] the filter parameters when the integrator is used in a higher order topology.

5.4.2 Experimental results

This section illustrates the performance of a second-order biquad, identical to that previously described in Section 5.3, fabricated with metal/poly capacitors whose values are collected in Table 5.6. The area of the input transistor was 8 μm^2. A microphotograph of the filter is shown in Fig. 5.26.

Figure 5.27 illustrates the THD versus the input amplitude measured in the lowpass filter for a single input tone of 10 kHz. A peak amplitude of 0.4 V results in less than 1 per cent THD. The HD3 is under 0.5 per cent. Figure 5.28 and Fig. 5.29 illustrate the tuning of different filter parameters. The cut-off frequency can only be changed in the range from 75 kHz to 120 kHz. The maximum possible cut-off frequency is lower than in the example in Section 5.3 due to both the smaller transconductance and the effect of the parasitic capacitances in the filter state space equations. The maximum achievable quality factor is around 2. This value is limited because of the degradation of the output resistance. The gain can be programmed in a 15 dB range. The resulting offset for the low pass filter is 75 mV, and additional frequency roots appear just half a

---

[63] Depending on how the filter frequency band compares with the values in eq. (5.118) and (5.119).

Table 5.6   *Input capacitances for the FGMOS transistors in a metal/poly realisation of Fig. 5.18 (fF)*

C_{VDDf} & C_{gndf}	126.2
C_{CMn} & C_{CMp}	37.5
C_i	11.15
C_{VDD}	47.62
C_{CM}	11.15

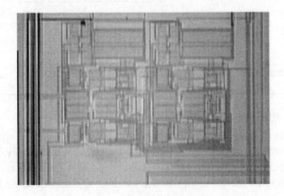

Figure 5.26   *Microphotograph of the second-order filter in Fig. 5.20, implemented with metal/poly capacitors*

decade from the nominal cut-off frequency. They can be observed in Fig. 5.28 where the slope of the high frequency rejection band is steeper than the ideal 20 dB/dec for the bandpass realisation. Also, the lowpass transfer function shows a zero at around 150 kHz for a filter with a cut-off frequency of 70 kHz. Regarding the low frequency losses of the bandpass filter, they are just under 20 dB below the peak value. This is related to the degradation of the gain in the transconductor due to mismatch. A summary of performance can be found in Table 5.7.

The main conclusion that can be extracted from these experimental results is that FGMOS circuits implemented in single poly technologies[64] can work, but their performance will be much poorer than the performance of the same circuit realised in a double poly technology. However, the latter will be equivalent if: (a) the area is not a constraint; (b) the technology is such that the parasitic capacitance between the FG and the bulk/well underneath is of the same order of magnitude as in a double poly technology; (c) the matching between capacitors implemented in the metal/poly technology is comparable to the matching in a double poly technology.

---

[64] With no high quality capacitors.

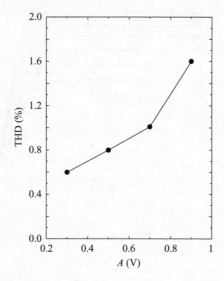

Figure 5.27 *Total Harmonic Distortion versus input amplitude for the filter in Fig. 5.20, implemented with metal/poly capacitors*

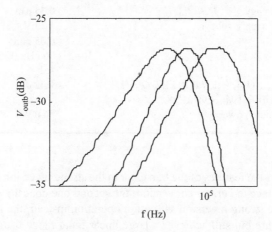

Figure 5.28 *Cut-off frequency tuning in the BP function for the filter in Fig. 5.20, implemented with metal/poly capacitors*

5.5 Comparison

The last two chapters have presented the design of three different OTA-C integrator/filter topologies, based on the FGMOS transistor operating in the strong inversion region. All the topologies have as common aims the reduction of the power consumption and power supply voltage while keeping a good circuit performance. The

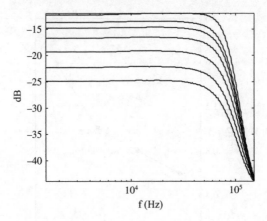

Figure 5.29   Gain tuning in the LP function for the filter in Fig. 5.20, implemented with metal/poly capacitors

Table 5.7   Summary of performance parameters for the filter in Fig. 5.20, implemented with metal/poly capacitors

Technology	0.35 μm
Power supply voltage	1.25 V
Area	0.05 mm^2
f_0	70 kHz–125 kHz
Q	<2
THD$_{max}$ (V_{pp} < 1 V@10 kHz, $Q=1$, $H_{LP}(0)=1, f_0 = 90$ kHz)	<42 dB
Maximum power	56 μW

first one employs the features of the transistor in the strong inversion ohmic region to linearise the voltage to current conversion; the second one does the same but taking advantage of the strong inversion saturation operation instead; the last one is not a linearised structure but still achieves a large linear input range thanks to the signal compression at the transistor FG. This topology also proves the feasibility of the device to operate at intermediate frequencies consuming very little power even with a low power supply voltage.

Table 5.8 compares the performance of the three main transconductor blocks in terms of the main design target parameters.

As the table shows the first two transconductors are suitable for low frequency applications. The second transconductor (transconductor based on FGMOS transistors biased in the strong inversion saturation region) requires a higher voltage supply to operate, offering however a larger programmability range than the first one (transconductor based on FGMOS transistors biased in the strong inversion ohmic region).

Table 5.8   Comparison between the three OTA realisations

	OTA Chapter 4	OTA Chapter 5 Section 5.2	OTA Chapter 5 Section 5.3
Power supply	1.5 V	2 V	1.25 V
Power consumption	6.5 μW (1 kHz)	8 μW (1 kHz)	181 μW (5 MHz)
Programming rate [G_m]	5.3 [11.9 ns, 62.8 ns]	100 [3 ns, 300 ns]	7.1 [6.6 μs, 46.7 μs]
Input range (THD < 1%)	2 V$_{pp}$	2 V$_{pp}$	1.2 V$_{pp}$

The third transconductor is more suitable for intermediate frequencies applications. Nevertheless, as it is the only non-linearised topology, the input range is smaller.

When comparing the values in the table, it should be kept in mind that the comparison between the two first and the third topologies is not totally fair, since the results shown for the last one correspond to a faster technology with lower threshold voltage values. This is also one of the reasons why, even when it operates at a frequency that it is almost three orders of magnitude higher than the maximum operating frequency of the other two, the power consumption is only around 20 times larger. Another reason is that in order to have a wide input range in the first two designs, the difference $V_{FG} - V_{Tn}$ has to be large enough in the middle of the range (to keep the transistors properly biased for low input signal values as well). This unavoidably increases the power consumption.

Notation

$\beta_{1(a,b)}$	β parameter of input transistors M1$_a$ and M1$_b$, respectively, in Fig. 5.18. Only different when there is mismatch
β_1	β parameter of input transistors in the transconductor in Fig. 5.1. Also, in Fig. 5.18 when $\beta_{1a} = \beta_{1b}$
β_i	Mi β parameter
γ_n	Parameter associated to the Flicker noise of n-channel transistors (Table 5.1)
γ_p	Parameter associated to the Flicker noise of p-channel transistors (Table 5.1)
Δ_{1i} for $i = [2, 4]$	Percentage of mismatch in capacitance C_{in} for transistor Mi (eq. (5.33). Transconductor in Fig. 5.1)
Δ_{2i} for $i = [2, 4]$	Percentage of mismatch in capacitance C_{CM} for transistor Mi (eq. (5.34). Transconductor in Fig. 5.1)
Δ_{3i}	Percentage of mismatch in capacitance C_{VDD} for transistor Mi (eq. (5.52). Transconductor in Fig. 5.1)
ω_o	Cut-off frequency

ω_{o1}	Coefficient of x_2 in the state space eq. (5.60) and of both x_1 and x_2 in the state space eq. (5.92)
ω_{o2}	Coefficient of x_1 in the state space eq. (5.60) and also in state space eq. (5.93)
ω_{o3}	Coefficient of x_1 in the state space eq. (5.61) and of the input in state space eq. (5.92)
ω_{o4}	Coefficient of the input (u) in the state space eq. (5.60)
ρ_n	Parameter associated to the thermal noise of n-channel transistors (Table 5.1)
ρ_p	Parameter associated to the thermal noise of p-channel transistors (Table 5.1)
(a,b)	Subscripts used to simultaneously refer to transistors $M1_a$ and $M1_b$, respectively
A_o	DC gain of the CMFB (Fig. 5.5, eq. (5.24))
a_0	See eq. (5.26). Value of a_{01} and a_{02} when both of them are the same
a_{0i} for $i = [1, 2]$	See eqs. (5.22) and (5.23)
A	Amplitude of a sinusoidal input signal
A'	See eq. (5.106)
A_{CM}	See eq. (5.78)
A_{CMa}	A_{CM} parameter for the transconductor branch named with subscripts a in Fig 5.18
A_{CMb}	A_{CM} parameter for the transconductor branch named with subscripts b in Fig. 5.18
A_d	Transconductor differential output gain (Fig. 5.18, eq. (5.83))
A_i	Area of the capacitance used to get a certain effective threshold voltage (eq. (5.99))
A_t	Area and perimeter of the transistor (eq. (5.99))
C	Integrating capacitance
C_c	Capacitance connected to either V_{b1} or V_{b2} (depending on the transistor) in the transconductor in Fig. 5.1
C_{CM}	Capacitance connected to the output of the CMFB in the transconductor input FGMOS transistors (Fig. 5.1 and Fig. 5.18)
C_{CMn}	Value of the capacitance connected to the output of the CMFB in the n-channel cascode transistor (Fig. 5.18)
C_{CMp}	Value of the capacitance connected to the output of the CMFB in the p-channel cascode transistor (Fig. 5.18)
$(C_{CM}/C_T)_a$	Input weight associated to capacitance C_{CM} in transistor $M1_a$ when there is mismatch (eq. (5.88). Transconductor in Fig. 5.18)
$(C_{CM}/C_T)_b$	Input weight associated to capacitance C_{CM} in transistor $M1_b$ when there is mismatch (eq. (5.88). Transconductor in Fig. 5.18)

$(C_{CM}/C_T)_i$	Input weight associated to capacitance C_{CM} in transistor Mi when there is mismatch (eq. (5.34). Transconductor in Fig. 5.1)
$(C_{CMn}/C_T)_a$	Input weight associated to capacitance C_{CMn} in transistor M2$_a$ when there is mismatch (eq. (5.86). Transconductor in Fig. 5.18)
$(C_{CMn}/C_T)_b$	Input weight associated to capacitance C_{CMn} in transistor M2$_b$ when there is mismatch (eq. (5.86). Transconductor in Fig. 5.18)
$(C_{CMp}/C_T)_a$	Input weight associated to capacitance C_{CMp} in transistor M3$_a$ when there is mismatch (eq. (5.86). Transconductor in Fig. 5.18)
$(C_{CMp}/C_T)_b$	Input weight associated to capacitance C_{CMp} in transistor M3$_b$ when there is mismatch (eq. (5.86). Transconductor in Fig. 5.18)
C_{GDi}	Gate to drain coupling for transistor Mi$_{(a,b)}$ in Fig. 5.18
C_{GSi}	Gate to source coupling for transistor Mi$_{(a,b)}$ in Fig. 5.18
C_{gnd}	Value of the capacitance connected to ground in the CMFB in Fig. 5.5 and in Fig. 5.19
C_{gndf}	Value of the capacitance connected to ground in the p-channel cascode transistor (Fig. 5.18)
$(C_{gnd}/C_T)_a$	Input weight associated to capacitance C_{gnd} in transistor M6$_a$ when there is mismatch (eq. (5.90). CMFB in Fig. 5.19)
$(C_{gnd}/C_T)_b$	Input weight associated to capacitance C_{gnd} in transistor M6$_b$ when there is mismatch (eq. (5.90). CMFB in Fig. 5.19)
C_i for $i = [1, N]$	Value of the capacitances connected to the effective inputs in the transconductor in Fig. 5.18 (when there is more than one)
C_{in}	Value of the effective inputs capacitances in the CMFBs (each one of them connected to a single output. Fig. 5.5 and Fig. 5.19) and in the input transistors of the transconductors in Fig. 5.1 (connected to V_{in1} and V_{in2}) and Fig. 5.18 (when the transconductor only has one effective input)
$(C_{in}/C_T)_i$	Input weight associated to capacitance C_{in} in transistor Mi when there is mismatch (eq. (5.33). Transconductor in Fig. 5.1)
$(C_{in}/C_T)_a$	Effective input weight in the transconductor branch named with subscripts a (Fig. 5.18) assuming mismatch
$(C_{in}/C_T)_b$	Effective input weight in the transconductor branch named with subscripts b (Fig. 5.18) assuming mismatch
C_{LF}	Load at the output of the CMFB (eq. (5.25))
C_{ox}	Capacitance per unit of area for a MOS transistor
C_{pa}	Parasitic capacitance at the output in branch a for the transconductor in Fig. 5.18
C_{pb}	Parasitic capacitance at the output in branch b for the transconductor in Fig. 5.18

C_{pma}	Poly1 to metal capacitance per unit of area
C_{pmp}	Poly1 to metal capacitance per unit of perimeter
C_{pwa}	Poly1 to well capacitance per unit of area
C_{pwp}	Poly1 to well capacitance per unit perimeter
C_T	Total capacitance seen by the FGs. It has been assumed the same in all the FGMOS transistors unless the opposite is said
C_{Ta}	Total area capacitance (eq. (5.100))
C_{Ti}	Total capacitance for transistor $Mi_{(a,b)}$ in Fig. 5.18 considering mismatch (metal-poly realisation)
C_{Tp}	Total perimeter capacitance (eq. (5.100))
C_{VDD}	Value of the capacitance connected to V_{DD} in the FGMOS input transistors (Transconductors in Fig. 5.1 and Fig. 5.18)
$(C_{VDD}/C_T)_a$	Input weight associated to capacitance C_{VDD} in transistor $M1_a$ when there is mismatch (eq. (5.86). Transconductor in Fig. 5.18)
$(C_{VDD}/C_T)_b$	Input weight associated to capacitance C_{VDD} in transistor $M1_b$ when there is mismatch (eq. (5.86). Transconductor in Fig. 5.18)
$(C_{VDD}/C_T)_i$	Input weight associated to capacitance C_{VDD} in transistor Mi when there is mismatch (Transconductor in Fig. 5.1 and CMFB in Fig. 5.5)
C_{VDD6}	Value of the capacitance connected to V_{DD} in the CMFB (Fig. 5.19) for the transconductor in Fig. 5.18
C_{VDDf}	Value of the capacitance connected to V_{DD} in the n-channel cascode transistor (Fig. 5.18)
$(C_{VDDf}/C_T)_a$	Input weight associated to capacitance C_{VDDf} in transistor $M2_a$ when there is mismatch (eq. (5.86). Transconductor in Fig. 5.18)
$(C_{VDDf}/C_T)_b$	Input weight associated to capacitance C_{VDDf} in transistor $M3_b$ when there is mismatch (eq. (5.86). Transconductor in Fig. 5.18)
del1	Percentage of mismatch between t_a and t_b (eq. (5.109))
$FG1_a$	FG in transistor $M1_a$ (Transconductor in Fig. 5.18)
$FG1_b$	FG in transistor $M1_b$ (Transconductor in Fig. 5.18)
f_o	Cut-off frequency in Hz
g_{dsi}	MOS transistor Mi output conductance
g_{dsFi}	Effective conductance for FGMOS device Mi
g_{mbi}	Bulk transconductance for transistor Mi
G_m	Transconductance of the transconductor in Fig. 5.1 (eq. (5.4))
G_{mi} for $i = [1, 4]$	Value of the transconductance that implements the coefficient ω_{oi}
g_{mi}	Transistor Mi transconductance
$G_{m(max)}$	Maximum value of the transconductance for the transconductor in Fig. 5.1 (eq. (5.5))

G_{outa}	Output conductance of the a branch in the transconductor in Fig. 5.18
G_{outb}	Output conductance of the b branch in the transconductor in Fig. 5.18
G_{outn}	Output conductance for the bottom (n-type) part of the integrator in Fig. 5.1 (eq. (5.17))
G_{outp}	Output conductance for the top (p-type) part of the integrator in Fig. 5.1 (eq. (5.18))
G_{outn}'	See eq. (5.27)
G_{outp}'	See eq. (5.28)
$H_{\text{BP}}(\omega_0)$	Bandpass filter gain at the cut-off frequency
$H_{\text{LP}}(0)$	Lowpass filter DC gain
I_1	Sum of M1 and M2 drain currents in the transconductor in Fig. 5.1
I_2	Sum of M3 and M4 drain currents in the transconductor in Fig. 5.1
I_{1a}	Drain current in transistor M1$_a$ (Transconductor in Fig. 5.18)
I_{1b}	Drain current in transistor M1$_b$ (Transconductor in Fig. 5.18)
I_B	Bias current provided by the load (eq. (5.2) and Fig. 5.1)
I_{Bmax}	Maximum value of the bias current (eq. (5.7))
I_{Bmin}	Minimum value of the bias current (eq. (5.8))
I_{out}	Output current in the transconductors (Fig. 5.1 and Fig. 5.18)
I_{SS}	Tail current in the transconductor in Fig. 5.18
k_1	See eq. (5.39)
k_{ca}	Parameter related to the intrinsic parasitic capacitances (eq. (5.100))
k_{cp}	Parameter related to the intrinsic parasitic capacitances (eq. (5.101))
$k_i A_i$	Area of the total designed capacitor used to process signals
$k_i P_i$	Perimeter of the total designed capacitor used to process signals
offset	Differential offset (eq. (5.107))
P	Power consumption
p	Pole of the CMFB (eq. (5.25) for the CMFB in Fig. 5.5 and eq. (5.111) for the CMFB in Fig. 5.18)
P_i	Perimeter of the capacitor used to get a certain effective threshold voltage (eq. (5.99))
Q	Quality factor
P_{min}	Minimum value of the power consumed by the transconductor in Fig. 5.1 (eq. (5.13))
P_{max}	Maximum value of the power consumed by the transconductor in Fig. 5.1 (eq. (5.13))

PSRR+	Positive power supply rejection ratio
P_t	Perimeter of the transistor (eq. (5.99))
R_{out}	Output resistance
R_{outa}	Output resistance of the a branch in the transconductor in Fig. 5.18
R_{outb}	Output resistance of the b branch in the transconductor in Fig. 5.18
S_i	Power spectral density of noise for transistor Mi
S_i'	$\sqrt{S_i}$
u	Input in the state space equation (5.60)
u_n	Single input (eq. (5.66))
u_p	Single input (eq. (5.66))
V_{cmN}	See eq. (5.105)
V_{1n}	Single voltage implementing x_{1n} in eq. (5.68)
V_{1p}	Single voltage implementing x_{1p} in eq. (5.68)
V_{2n}	Single voltage implementing x_{2n} in eq. (5.69)
V_{2p}	Single voltage implementing x_{2p} in eq. (5.69)
V_{BP}	Bandpass output (eq. (5.68))
V_b	Bias voltage for zero transconductance in the transconductor of Fig. 5.1 (eq. (5.8))
V_{b1}	Bias voltage for the integrator in Fig. 5.1
V_{b1min}	V_{b1} minimum value (eq. (5.5))
V_{b2}	Bias voltage for the integrator in Fig. 5.1
V_{b2max}	V_{b2} maximum value (eq. (5.5))
V_{casn}	Bias voltage for the n-channel cascode transistor in the integrator in Fig. 5.1
V_{casp}	Bias voltage for the p-channel cascode transistor in the integrator in Fig. 5.1
V'_{CM}	$V_{inCM} + (C_{CM}/C_{in})V_{outcm} + (C_C/C_{in})V_{b2}$
$V_{CM,min}$	Minimum value of V'_{CM}, in Fig. 5.9
$V_{CM,max}$	Maximum value of V'_{CM}, in Fig. 5.9
V_{D1}	M1 drain voltage for the transconductor in Fig. 5.1. Also, M1a and M1b drain voltages when common mode signals are applied and both transistors are assumed to be identical for the transconductor in Fig. 5.18
V_{D1min}	Minimum M1 drain voltage for the transconductor in Fig. 5.1
V_{D1a}	M1$_a$ drain voltage in Fig. 5.18
V_{D2}	M2$_a$ and M2$_b$ drain voltages when common mode signals are applied and both transistors are assumed to be identical
V_{D2a}	M2$_a$ drain voltage in Fig. 5.18. It is also one of the single output voltages
V_{D3}	M3 drain voltage in the transconductor in Fig. 5.1
V_{D12}	M12 drain voltage (Fig. 5.5)
$V_{i(a,b)}$ for $i = [1, N]$	Effective inputs in the transconductor in Fig. 5.18

V_{iCM}	Common mode at the input of the transconductor in Fig. 5.18
V_{id}	Differential input voltage for the transconductor in Fig. 5.18
V_{inn}	Single voltage implementing u_n in eq. (5.68)
V_{in1}	Single input voltage for the transconductor in Fig. 5.1
V_{in2}	Single input voltage for the transconductor in Fig. 5.1
$V_{in(a,b)}$	Effective inputs in branches a and b, for the transconductor in Fig. 5.18 when only one input is considered
V_{ind}	Differential input for the integrator in Fig. 5.1 and for the transconductor in Fig. 5.18 when only one effective input is considered
V_{inCM}	Common mode at the input of the integrator in Fig. 5.1. Also, common mode at the effective input of the transconductor in Fig. 5.18 when only one input is considered
V_{inRMS}	RMS spectral density of noise at the effective input for the transconductor in Fig. 5.1
$\dfrac{\overline{v_{in}^2}}{\Delta f}$	Equivalent power spectral density of noise at the effective input
V_{inp}	Single voltage implementing u_p in eq. (5.68)
V_{LP}	Lowpass output (eq. (5.69))
V_{off+}	Equivalent single side offset (eq. (5.104))
V_{out}	Differential output voltage for the transconductors in Fig. 5.1 and Fig. 5.18
V_{outCM}	Common mode at the output of the transconductor in Fig. 5.1 and in Fig. 5.18
V_{outcm}	Output of the CMFB
$V_{outcm(max)}$	Maximum V_{outcm} (eq. (5.7))
$V_{outcm(min)}$	Minimum V_{outcm} (eq. (5.7))
V_{out1}	Single output voltage for the transconductor in Fig. 5.1
$\dfrac{\overline{\Delta v_{out1}^2}}{\Delta f}$	Equivalent power spectral density of noise at the effective input of the CMFB used for the transconductor in Fig. 5.1 (Fig. 5.5)
V_{out2}	Single output voltage for the transconductor in Fig. 5.1
V_{outb}	Bandpass output in the filter in Fig. 5.20
V_{outl}	Lowpass output in the filter in Fig. 5.20
V_S	M12 and M13 source voltage (Fig. 5.11)
V_{S1}	Input transistors $M1_{(a,b)}$ source voltage (Transconductor in Fig. 5.18)
V_{S3}	$M3_a$ and $M3_b$ source voltages when common mode signals are applied and both transistors are assumed to be identical (Fig. 5.18)

V_{S3a}	$M3_a$ source voltage in Fig. 5.18
V_{S8}	M8 source voltage (Transconductor in Fig. 5.1)
V_{TFG}	Effective threshold voltage (eq. (5.9) and eq. (5.75)) (Different values for different transistors)
V_{Tn}	n-channel MOS transistors threshold voltage (it has been assumed the same in all the devices, unless the opposite is said)
V_{Tn1}	M1 threshold voltage considering mismatch and body effect (eq. (5.103), Fig. 5.18)
V_{Tn2}	M2 threshold voltage considering mismatch and body effect (eq. (5.103), Fig. 5.18)
$|V_{TP}|$	p-channel MOS transistors threshold voltage (it has been assumed the same in all the devices, unless the opposite is said)
w	w_i when all the weights for the effective inputs have been assumed to have the same value
w_i	C_i/C_T. Effective inputs weight (Transconductor in Fig. 5.18)
x_1	Bandpass output
x_{1n}	Single state variable in eq. (5.66). (Related to the bandpass output)
x_{1p}	Single state variable in eq. (5.66). (Related to the bandpass output)
x_2	Lowpass output
x_{2n}	Single state variable in eq. (5.66). (Related to the lowpass output)
x_{2p}	Single state variable in eq. (5.66). (Related to the lowpass output)
x_i	Input in the state space equation (5.92)

Chapter 6

Low power analog continuous-time G_m-C filtering using the FGMOS in the weak inversion region

The following two chapters are devoted to exploring the potential of the FGMOS transistor operating in weak inversion for the design of analog low-voltage micropower continuous-time filters. It is demonstrated that the use of FGMOS in weak inversion as an alternative to conventional MOS devices relaxes the constraints relating to voltage supply and frequency response. Moreover, the increased number of terminals in the device permits a simplification in the topologies required to realise a certain mathematical function. This brings with it a reduction in power consumption, as well as other advantages such as a lower noise floor.

The term 'weak inversion' defines the operating mode in an MOS transistor in which the bulk surface charge is inverted in relation to the rest of it, but still it is mostly due to the charge in the depletion region. The corresponding inversion layer charge can nevertheless cause non-negligible conduction. Weak inversion mode is especially suited for the design of micropower circuits for several reasons. The term 'micropower' defines a class of circuits with power consumption of just a few microwatts. In order to obtain these power levels the maximum current the circuit can handle has to be very small. The weak inversion region is very convenient when these requirements have to be met for the following reasons:

1. Assuming that the required power consumption sets a very small, fixed value for the current, the maximum operating frequency of a single transistor is determined by the gate oxide capacitances, C_{GB}, C_{GS} and C_{GD}. In order to maximise the device bandwidth the latter needs to be kept as small as possible, which implies a small device area. Assuming that the minimum transistor width is chosen, the designer must choose a length of transistor and level of inversion in the channel that meets the current constraint. If the designer reduces the level of inversion, the length of the transistor may also be reduced. This reduces C_{GB}, C_{GS} and C_{GD}, which in turn increases the bandwidth.

2. The maximum voltage differences between the terminals of a device working in the weak inversion region are smaller than the values demanded by strong inversion operation. This allows for lower supply voltages, which also reduces the power consumption.

3. The behaviour of the transistor is ruled by an exponential law whose derivative is another exponential function. This makes it possible to use nonlinear processing internally to implement linear state-space equations, in a simple form [182]. By doing this, the internal nodal voltage swings can be reduced and less power is needed to charge and discharge the nodes.

This chapter describes a linearised G_m-C topology in which the linearisation is achieved by means of two 'floating gate' blocks: a nonlinear transconductor with three output currents and a square-root block based on FGMOS transistors biased in the weak inversion region that processes the three currents provided by the first block, thus giving rise to a fully linear topology.

The chapter begins with a reformulation of the Translinear Principle (TP), which is an ideal technique for implementing nonlinear functions using currents in FGMOS. The advantages of an FGMOS implementation compared with a normal MOS implementation in a low-voltage context are then discussed. The linearised G_m-C integrator is described afterwards followed by an analysis of second-order effects, such as PSRR or THD, that can affect its performance. Finally the design of a second-order filter based on these blocks is explained and experimental results are shown to illustrate the conclusions.

6.1   FGMOS translinear principle

In 1975, B. Gilbert applied the name 'translinear' to an emerging class of circuits whose large-signal behaviour was based on the exponential current–voltage characteristic of the bipolar transistor and the linear voltage law of a voltage loop [218]. By combining both principles, a nonlinear product of current densities within different loops can be obtained that serves as a basis for many nonlinear circuits, including wideband analog multipliers, RMS-DC converters and vector-magnitude circuits [218–222]. The general circuit principle describing the specific topological arrangement for the exponential devices which gives rise to the desired mathematical function is called the Translinear Principle (TP). This section explains how the TP can be used advantageously when FGMOS devices are available.

Consider two FGMOS transistors M0 and M1 with a grounded source and a common input, V_{0i} at M0 and V_{1j} at M1, as illustrated in Fig. 6.1. M0 has N and M1 has M inputs. The Kirchoff's voltage law in this trivial loop is

$$-V_{0i} + V_{1j} = 0 \tag{6.1}$$

Under the assumptions that the parasitics to the FG are very small compared with the input capacitances and that the charge trapped during the fabrication process is negligible, the drain current in FGMOS devices working in weak inversion saturation

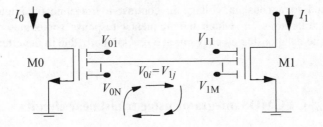

Figure 6.1   FGMOS translinear loop

mode with their source and bulk grounded is given by (see eq. (2.5))

$$I_D = I_s e^{\sum_{i=1} w_i V_i / n U_T} \tag{6.2}$$

where w_i refers to the input weights and V_i to the inputs. Expressing the voltages in the translinear loop as a function of the currents flowing through the transistors gives

$$V_{0i} = nU_T \left[\ln \left(\frac{I_0}{I_S} \right)^{1/w_{0i}} - \frac{1}{w_{0i}} \times \sum_{\substack{k=1 \\ k \neq i}}^{N} \frac{w_{0k} V_{0k}}{n U_T} \right]$$

$$V_{1j} = nU_T \left[\ln \left(\frac{I_1}{I_S} \right)^{1/w_{1j}} - \frac{1}{w_{1j}} \times \sum_{\substack{k=1 \\ k \neq j}}^{M} \frac{w_{1k} V_{1k}}{n U_T} \right] \tag{6.3}$$

where w_{0k}, w_{1k} are the equivalent weights for the input k in transistors M0 and M1, respectively. Rewriting (6.1) using (6.3) and rearranging the terms gives

$$\left(\frac{I_0}{I_s} \right)^{1/w_{0i}} e^{-\sum_{\substack{k=1 \\ k \neq i}}^{N} (w_{0k} V_{0k}/w_{0i} n U_T)} = \left(\frac{I_1}{I_S} \right)^{1/w_{1j}} e^{-\sum_{\substack{k=1 \\ k \neq j}}^{M} (w_{1k} V_{1k}/w_{1j} n U_T)} \tag{6.4}$$

This is the nonlinear relationship between currents, which is the objective of the TP. The first advantage of using the FGMOS to implement the translinear loops is that it is possible to realise the exponents in the current function with capacitance ratios, and hence the source terminal can be fixed to a constant voltage. This fact is quite significant for two different reasons. The first is that the voltage requirements are reduced since only a V_{GS} (gate to source voltage drop) is needed in the loop and therefore there is no need to stack transistors. The second is that the lack of internal nodes both simplifies and improves the frequency response and eliminates instabilities in certain circuit topologies. This is because the C_{GS} and C_{GB} capacitors are now tied to a constant voltage in contrast to translinear circuits designed with MOS transistors, in which the frequency response sometimes exhibits an undesired and even unstable behaviour at a few kilohertz due to the effects of these parasitics.

6.2 A G_m-C FGMOS integrator using translinear circuits

This section shows how to use the FGMOS transistors as circuit primitives to design linear and nonlinear (square-root) circuits. Once more lower voltage operation is obtained by shifting the voltage levels wherever needed. However, for the first time now in this book, the FGMOS is presented as a tool to realise nonlinear functions through the TP. Combining the two saturated modes in the transistor (in strong inversion and weak inversion), a design flow is developed that generates an FGMOS transistor based circuit from a set of state space equations.

6.2.1 The state-space integrator equations

The starting point of the filter design process is a set of first-order differential equations of the type

$$\dot{x} + \alpha x = \eta x_i \tag{6.5}$$

where x is a state variable; x_i is the external excitation which can be an independent source, other state variable or any combination of both; and parameters α and η are related to the filter specifications. The fundamental idea behind the proposed implementation relies on the versatile operation of the FGMOS technique which permits the realisation of complex algebraic operations under low voltage conditions. In order to make the filter less sensitive to noise and enlarge the linear range at the input, a fully differential operation is chosen. Hence, equation (6.5) has to be rewritten as

$$(\dot{x}_p - \dot{x}_n) = -\alpha(x_p - x_n) + \eta(x_{ip} - x_{in}) \tag{6.6}$$

Assuming that all these signals are voltages that have been obtained from current signals, the following variable changes are introduced:

$$V_p - V_n = (I_p - I_n)/(K_1\sqrt{I_c}) \tag{6.7}$$

$$V_{ip} - V_{in} = (I_{ip} - I_{in})/(K_1\sqrt{I_{ic}}) \tag{6.8}$$

where $I_c, I_p, I_n, I_{ic}, I_{ip}, I_{in}$ are the currents and K_1 is a constant. These parameters are discussed in detail later.

Using eq. (6.7) and eq. (6.8) and applying the variable change to the right-hand side (RHS) of eq. (6.6), the final differential equation is obtained:

$$(\dot{V}_p - \dot{V}_n) = -\alpha\frac{I_p - I_n}{K_1\sqrt{I_c}} + \eta\left(\frac{I_{ip} - I_{in}}{K_1\sqrt{I_{ic}}}\right) \tag{6.9}$$

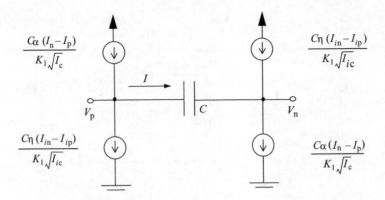

Figure 6.2   Schematic of an ideal circuit that performs the first-order state-space eq. (6.10)

Then, multiplying both sides in eq. (6.9) by C (physically an integrating capacitor) allows the RHS of eq. (6.9) to be identified with the current flowing through this capacitor C:

$$I = C(\dot{V}_p - \dot{V}_n) = -C\alpha \frac{I_p - I_n}{K_1\sqrt{I_c}} + C\eta \left(\frac{I_{ip} - I_{in}}{K_1\sqrt{I_{ic}}} \right) \qquad (6.10)$$

Equation (6.10) is symbolised in Fig. 6.2 using ideal circuit elements. Referring to Fig. 6.2, it can be seen that Kirchoff's current law is evaluated in nodes V_p and V_n according to eq. (6.10).

6.2.2   FGMOS blocks

6.2.2.1   Nonlinear transconductor

A circuit that performs the current to voltage conversion required by equations (6.7) and (6.8) is shown in Fig. 6.3. It operates as follows:

If the three FGMOS transistors are equally sized, operate in the strong inversion saturation region and have the same input weights, given by

$$w_{VDD} = C_{VDD}/C_T = w_{DD}, \quad w_{in} = C_{in}/C_T = w/2 \qquad (6.11)$$

the currents flowing through them will be

$$I_p \approx \frac{\beta_n}{2}(w_{DD}V_{DD} + wV_p - V_{Tn})^2 \quad I_n \approx \frac{\beta_n}{2}(w_{DD}V_{DD} + wV_n - V_{Tn})^2 \qquad (6.12)$$

$$I_c \approx \frac{\beta_n}{2}\left(w_{DD}V_{DD} + \frac{w}{2}(V_p + V_n) - V_{Tn}\right)^2 \qquad (6.13)$$

where parameters β_n and V_{Tn} have their usual meanings, and second-order effects such as the parasitics to the FG, the possible residual charge trapped in the interface oxide-silicon and the non-zero output conductance, have been neglected. If all the constant

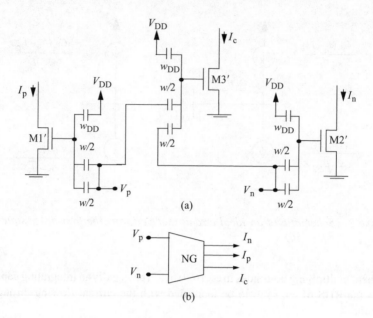

Figure 6.3   (a) FGMOS circuit for variable change. (b) Symbol

terms in the previous equations are grouped together, a new effective threshold voltage can be defined and is given by

$$V_{TFG} = V_{Tn} - w_{DD}V_{DD} \tag{6.14}$$

Equation (6.14) shows a reduced value of the threshold voltage obtained by connecting one of the inputs to the maximum voltage available in the circuit (V_{DD}). This allows the operation of the transistor in the strong inversion region even under a tight constraint of low supply voltage. Taking the difference between I_p and I_n, it can be shown that the relationship between the differential voltage and current has the desired general form of eq. (6.7):

$$(V_p - V_n) = (I_p - I_n)/(w\sqrt{2\beta_n I_c}) \tag{6.15}$$

where I_c is defined as the common mode current derived from V_p and V_n. I_c senses the common mode voltage, $V_{cm} = (V_p + V_n)/2$, of both voltage signals. Equation (6.15) represents a nonlinear relationship between voltages and currents. From now on, we will refer to the circuit that realises eq. (6.15) as the nonlinear transconductor (NG-circuit), and its symbol is shown in Fig. 6.3(b). Each NG-circuit has a transconductance given by

$$G_{m(NG)} = w\sqrt{2\beta_n I_c} \tag{6.16}$$

6.2.2.2   The $y/\sqrt{x}$ circuit

Now, the RHS of eq. (6.9) can be easily implemented adding/subtracting currents in a summing node. The circuit required by eq. (6.15) has to realise a $y/\sqrt{x}$-type operator, which can be obtained using the TP previously depicted for the FGMOS in weak inversion saturation. Each RHS term in eq. (6.9) can be rewritten in a more general form as

$$\alpha I_{\rm d}/(K_1\sqrt{I_{\rm c}}) \tag{6.17}$$

$$\eta I_{id}/(K_1\sqrt{I_{ic}}) \tag{6.18}$$

where $K_1 = w\sqrt{2\beta_{\rm n}}$, $I_{\rm d} = I_{\rm p} - I_{\rm n}$ and $I_{id} = I_{ip} - I_{in}$. Multiplying both sides in eq. (6.9) by a capacitance C will give a dimensionally correct equality if

$$\frac{C\alpha}{K_1} = \frac{C\alpha}{w\sqrt{2\beta_{\rm n}}} = A_j\sqrt{I_{Aj}} \qquad \frac{C\eta}{K_1} = \frac{C\eta}{w\sqrt{2\beta_{\rm n}}} = A_i\sqrt{I_{Ai}} \tag{6.19}$$

where A_j and A_i are non-dimensional factors and I_{Aj} and I_{Ai} are independent currents. An alternative form of (6.10) is thus obtained:

$$C(\dot{V}_{\rm p} - \dot{V}_{\rm n}) = -A_j\sqrt{I_{Aj}}\frac{I_{\rm d}}{\sqrt{I_{\rm c}}} + A_i\sqrt{I_{Ai}}\frac{I_{id}}{\sqrt{I_{ic}}} \tag{6.20}$$

Two currents are required to implement eq. (6.20), one for the α coefficient and another for η. A circuit whose output has the general form of one of the RHS terms in eq. (6.20) is shown in Fig. 6.4(a). In eq. (6.20), $I_{\rm d}$ is any differential current (as for example $(I_{\rm p} - I_{\rm n})$) associated with a state variable, $I_{\rm c}$ is its associated common mode current and I_{Aj} is an independent current source. The independent voltage sources are used for shifting the threshold voltage and controlling the circuit gain. All the FGMOS transistors have to be working in the weak inversion saturation region. The corresponding single output currents for the circuit are now given by[65]

$$I_{o(n,p)} = A_j\sqrt{I_{Aj}/I_{\rm c}} \times I_{(n,p)} \tag{6.21}$$

where A_j depends on the constant reference voltage sources at the inputs of the FGMOS transistors. Considering the same total capacitance and sizes for all the FGMOS, the input capacitors are chosen to give the weights in Fig. 6.4, and all the extra inputs are set to a value $V_{\rm c}$ given by an independent voltage source (except V_{Aj}):

$$A_j = {\rm e}^{(V_{Aj}-V_{\rm c})/4nU_{\rm T}} \tag{6.22}$$

---

[65] Subscripts (n,p) in $I_{o(n,p)}$ have been used to refer to either I_{on} or I_{op}, and the same applies for $I_{(n,p)}$. This will apply to other variables from now on.

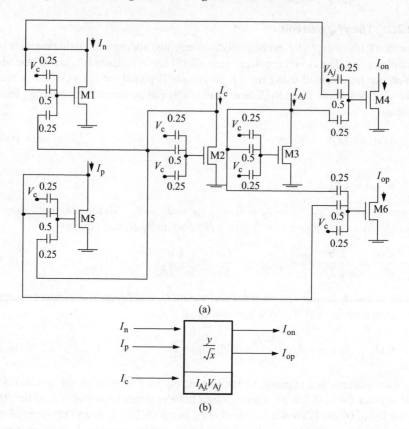

(b)

Figure 6.4 (a) $y/\sqrt{x}$ circuit with FGMOS-based TL loops. (b) Symbol

V_c is chosen to ensure weak inversion operation. V_{Aj} and I_{Aj} are used for tuning. Hence

$$\alpha = (w\sqrt{I_{Aj}}\sqrt{2\beta_n})e^{(V_{Aj}-V_c)/4nU_T}/C$$

$$\eta = (w\sqrt{I_{Ai}}\sqrt{2\beta_n}) \cdot e^{(V_{Ai}-V_c)/4nU_T}/C \qquad (6.23)$$

Negative values of $I_{o(n,p)}$ can be easily implemented with current mirrors. The symbol for the circuit in Fig. 6.4(a) is shown in Fig. 6.4(b).

6.2.2.3 The transconductor

The circuit that implements each term in the space state equation is shown in Fig. 6.5. From now on it will be called FG-block. The transconductance of this cell is given by

$$G_m = C/\tau = wA_j\sqrt{2k_e\beta_n I_{Aj}} \qquad (6.24)$$

where k_e is a constant that scales the magnitude of the currents coming from the NG-blocks. It is set by the ratio between the sizes of the output and input transistors

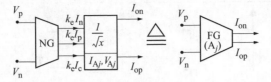

Figure 6.5   The FGMOS transconductor (FG)

in a pMOS current mirror that provides the input to the square-root block. The fact that this is a large signal transconductor makes it very useful in the context of LV designs, since having a linearised structure increases the input range for a given value of distortion, which is one of the critical factors as the power supply voltage is scaled down.

6.2.2.4   The state space circuit

The final realisation of the circuit that implements eq. (6.6) is shown in Fig. 6.6(c). Two equal G-blocks have been used, one for each voltage-node: the state variable x and the input voltage x_i. The G-block drawn in Fig. 6.6(a) and (b) is just an FG-block with an added cascode output stage that provides the differential currents. It can be shown that the use of cascode FGMOS transistors improves the output resistance (with all the consequences that this fact has on the frequency response of a higher-order filter designed with it). Also, the output swing is enlarged and the common mode feedback circuitry (required due to the fully differential structure) is simplified. One of the inputs to each cascode transistor is connected to the output of the common mode feedback block (CMFB) shown in Fig. 6.7. This CMFB is identical to the one described in Chapter 4, so the reader is referred to this chapter for more information about it.

6.2.3   Second-order effects

The currents injected into the integrating capacitors could deviate from their ideal behaviour mainly due to the following reasons: (a) mismatch between the input capacitances, and the effects of the parasitics (C_{GD}); (b) mismatch between the transistors (β_n and V_{Tn}); (c) output resistances (degradation of the square and exponential laws) and (d) dependence of the β parameter on gate voltage (θ). This section focuses on the analysis of how these effects can affect the performance of the transconductor.

6.2.3.1   Offset and nonlinearities

1. *Mismatch between the input capacitances and effect of the gate to drain coupling in diode-connected transistors.* If mismatch between the input capacitances is taken into account the currents given by eqs. (6.12) and (6.13) have to be rewritten as

$$I_{(n,p)} = \frac{\beta_n}{2}\left[w(1 + \Delta'_{(2,1)})\left(V_{cm} \pm \frac{V_d}{2}\right) - V_{TFG(n,p)}\right]^2 \qquad (6.25)$$

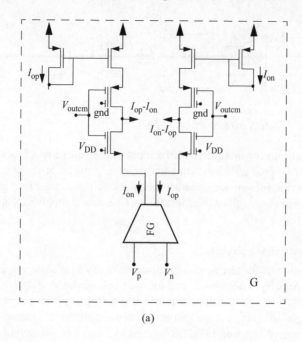

(a)

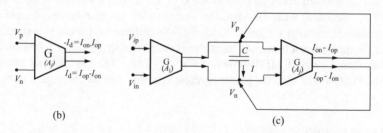

(b)   (c)

Figure 6.6   Differential transconductor. (a) G-block. (b) Symbol. (c) Implementation of eq. (6.6)

$$I_c = \frac{\beta_n}{2}\left[\frac{w}{2}(1 + \Delta'_{31})V_p + \frac{w}{2}(1 + \Delta'_{32})V_n - V_{TFGc}\right]^2 \qquad (6.26)$$

where the Δ'_i parameters represent the percentage of mismatch, and V_d is the difference between V_p and V_n. These expressions also take into account the deviations of the effective threshold voltages with respect to their nominal values. Now

$$V_{TFG(n,p)} = V_{Tn(n,p)} - w_{DD(2,1)}V_{DD} \qquad V_{TFGc} = V_{Tn} - w_{DD}V_{DD} \qquad (6.27)$$

In the square-root block, the influence of mismatch is demonstrated through the exponents of the currents in the nonlinear functions since they differ from the expected ideal values. However, there is another source of variation in the exponents with respect to their ideal values: the gate-to-drain parasitic coupling in those transistors

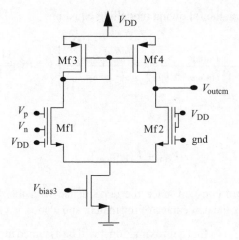

Figure 6.7   Common Mode Feedback Block (CMFB)

which have one of their inputs connected to drain. Both sources of error can be studied together since the effect they have in the final response is qualitatively the same. The general expression for the weight w, when all these deviations are taken into account is

$$w_{Ri} = w(1 + c_i) = w\left(1 + \Delta_i + \frac{C_{GDi}}{wC_{Ti}}\right) \tag{6.28}$$

And the real function implemented by the $1/\sqrt{x}$ circuit is given by

$$I_{o(n,p)} \approx A_{j(n,p)}I_s \left(\frac{I_{(n,p)}}{I_s}\right)^{1/(1+c_{1(n,p)})} \left(\frac{I_s}{I_c}\right)^{0.5/(1+c_{1(n,p)})(1+c_2)} \left(\frac{I_{Aj}}{I_s}\right)^{0.5/(1+c_3)}$$

$$= A'_{j(n,p)}\frac{I_{n,p}^{1+x'_{1(n,p)}}}{\sqrt{I_c}} \cdot I_c^{x'_{2(n,p)}} \tag{6.29}$$

2. *Mismatch between the transistors.* Mismatch between the transistors will cause a different multiplicative constant for the currents which can be included in the value of $A'_{j(n,p)}$.

3. *Output resistance.* The main contribution to degradations in the output resistance in the FGMOS is the capacitive coupling between the drain and the gate. For the diode connected transistors, it has been analysed previously in eq. (6.28). In the case of output transistors, their output resistances have been improved by cascoding, hence their influence will be negligible.

4. *Dependence of the β parameter on the gate voltage.* The dependence of the β parameter on the voltage at the FG of the transistors generating I_n and I_p can be taken

into account by using the following modelling equation:

$$\beta'_{n(n,p)} \approx \frac{\beta_{n(n,p)}}{1 + \theta(V_{GS(n,p)} - V_{Tn})} \approx \frac{\beta_{n(n,p)}}{(1 + k_{2(n,p)})}\left[1 \mp \theta w \frac{(1 + \Delta'_{(2,1)})}{2(1 + k_{2(n,p)})}V_d\right]$$

(6.30)

$$k_{2(n,p)} = \theta[w(1 + \Delta'_{(2,1)})V_{cm} - V_{TFG(n,p)}]$$

(6.31)

I_c can be considered constant since the dependence with the differential input is weak (it is scaled by the percentage of mismatch, and also x'_2 is very small compared with 0.5). Therefore, as a first approximation it will be assumed that $I_c^{x'_2} = I_c^{x'_{2n}} = I_c^{x'_{2p}}$. Also, the β of the transistor generating I_c will be referred as β_{nc}.

Taking all the previous considerations into account, and after a few laborious derivations, the following conclusions can be drawn:

1. Mismatch causes an offset, whose value can be calculated using the previously derived expressions, and is approximately

$$\text{offset} \approx \left(\frac{k_e\beta_{nn}}{2(1 + k_{2n})}a_{0n} - \frac{k_e\beta_{np}}{2(1 + k_{2p})}a_{0p}\right)\frac{I_c^{x'_2}}{\sqrt{k_e\beta_{nc}/2}}\varepsilon^{-1} + H_{D2}$$

(6.32)

The magnitude of this offset depends upon the common mode of the signal through ε. The variables that have not been defined previously are given in Table 6.1.

2. There is variation in the gain of the block. The new gain is

$$H_{D1} \approx \varepsilon^{-1}\frac{AI_c^{x'_2}}{\sqrt{k_e\beta_{nc}/2}}\left(\frac{k_e a_{1n}\beta_{nn} - k_e a_{1p}\beta_{np}}{2}\right) - \varepsilon^{-1}\frac{AI_c^{x'_2}}{\sqrt{k_e\beta_{nc}/2}}$$

$$\times \frac{\theta w a_{0n}k_e\beta_{nn}}{4(1 + k_{2n})}(1 + \Delta'_2) - \varepsilon^{-1}\frac{AI_c^{x'_2}}{\sqrt{k_e\beta_{nc}/2}}\frac{\theta w a_{0p}k_e\beta_{np}}{4(1 + k_{2p})}(1 + \Delta'_1)$$

$$- \frac{w(\Delta'_{31} - \Delta'_{32})}{4\varepsilon^2}\frac{AI_c^{x'_2}}{\sqrt{k_e\beta_{nc}/2}}\left(\frac{a_{0n}k_e\beta_{nn} - a_{0p}k_e\beta_{np}}{2}\right) + 3H_{D3}$$

(6.33)

where A represents the amplitude of a sinusoidal input signal.

Table 6.1   *Definition of the parameters used for the calculation of second-order effects*

ε	$w\left(1 + \dfrac{\Delta'_{32} + \Delta'_{31}}{2}\right)V_{cm} - V_{TFGc}$
$A_{0(n,p)} = a_{o(n,p)}$	$k_{1(n,p)}\left[1 + x'_1 \ln k_{1(n,p)} + \dfrac{x'^2_1}{2}(\ln k_{1(n,p)})^2 + \dfrac{x'^3_1}{6}(\ln k_{1(n,p)})^3 + \cdots\right]$
$A_{1(n,p)}$	$1 + x'_1 + x'_1 \ln k_{1(n,p)} + \dfrac{x'^2_1}{2} \ln k_{1(n,p)}$
	$\times \left(2 + \ln k_{1(n,p)}\left(1 + x'_1 + \dfrac{x'_1 \ln k_{1(n,p)}}{3}\right)\right)$
$a_{1(n,p)}$	$A_{1(n,p)}b_{1(n,p)}$
$A_{2(n,p)}$	$\dfrac{x'_1}{2k_{1(n,p)}} + \dfrac{x'^2_1}{2k_{1(n,p)}}(\ln k_{1(n,p)} + 1) + \dfrac{x'^3_1}{4k_{1(n,p)}} \ln k_{1(n,p)}(\ln k_{1(n,p)} + 2)$
$a_{2(n,p)}$	$A_{1(n,p)}B'_{(n,p)} + A_{2(n,p)}b^2_{1(n,p)}$
$A_{3(n,p)}$	$-\dfrac{x'_1}{6k^2_{1(n,p)}} - \dfrac{x'^2_1}{6k^2_{1(n,p)}} \ln k_{1(n,p)} + \dfrac{x'^3_1}{6k^2_{1(n,p)}}\left(1 - \dfrac{\ln k^2_{1(n,p)}}{2}\right)$
$a_{3(n,p)}$	$\dfrac{A_{2(n,p)}}{2}b_{1(n,p)}B'_{(n,p)} + A_{3(n,p)}b^3_{1(n,p)}$
$A'_{j(n,p)}$	$A_{j(n,p)}I_s^{[((c_{1(n,p)}(1+c_2)+0.5)/(I+c_{1(n,p)}))-0.5/(1+c_3)]}$
$B'_{(n,p)}$	$A'_{j(n,p)}\dfrac{w^2[1 + (\Delta'_{32} + \Delta'_{31})/2]^2}{4}E^2_{(n,p)}$
$b_{1(n,p)}$	$2C'_{(n,p)}\varepsilon + 2D'_{(n,p)}$
$c_{i(n,p)}$	$\left[\Delta_{i(n,p)} + C_{GDi(n,p)}/\left(wC_{T_{i(n,p)}}\right)\right]$
$C'_{(n,p)}$	$(\mp A'_{j(n,p)})(E^2_{(n,p)} \cdot w[1 + (\Delta'_{32} + \Delta'_{31})/2]/2)$
$D'_{(n,p)}$	$(\pm[A'_{j(n,p)}E^2_{(n,p)}(\Delta V_{TFG(n,p)}) \cdot w(1 + (\Delta'_{32} + \Delta'_{31})/2)])/2$
$E_{(n,p)}$	$(1 + \Delta'_{(2,1)})/[1 + (\Delta'_{31} + \Delta'_{32})/2]$
$= \sqrt{A'_{(n,p)}/A'_{j(n,p)}}$	
$F'_{(n,p)}$	$A'_{j(n,p)}E^2_{(n,p)}(\Delta V_{TFG(n,p)})^2$
$G'_{(n,p)}$	$-A'_{j(n,p)}E^2_{(n,p)}(\Delta V_{TFG(n,p)})$
$k_{1(n,p)}$	$A'_{(n,p)}\varepsilon^2 + 2G'_{(n,p)}\varepsilon + F_{(n,p)}$
$x'_{1(n,p)}$	$(-c_{1(n,p)}/(1 + c_{1(n,p)}))$
$x'_{2(n,p)}$	$0.5(c_{1(n,p)} + c_2 + c_{1(n,p)}c_2)/[(1 + c_{1(n,p)})(1 + c_2)]$
$\Delta V_{TFG(n,p)}$	$\dfrac{V_{TFG(n,p)}}{E_{(n,p)}} - V_{TFGc}$

3. The second-order harmonic should be zero, since the structure is fully differential. However, due to the mismatch, it is now

$$
H_{D2} \approx \left(-\frac{A^2 I_c^{x_2'}}{2\sqrt{k_e \beta_{nc}/2}} \right) \left(\frac{w(\Delta_{31}' - \Delta_{32}')}{4\varepsilon^2} \right)
$$

$$
\times \left(\frac{a_{1n}k_e\beta_{nn} - a_{1p}k_e\beta_{np}}{2} \right) + \frac{A^2 \theta w I_c^{x_2'}}{2\sqrt{k_e\beta_{nc}/2}} \left(\frac{w(\Delta_{31}' - \Delta_{32}')}{4\varepsilon^2} \right)
$$

$$
\times \left(\frac{a_{0n}k_e\beta_{nn}(1 + \Delta_2')}{4(1 + k_{2n})} + \frac{a_{0p}k_e\beta_{np}(1 + \Delta_1')}{4(1 + k_{2p})} \right)
$$

$$
+ \frac{A^2 I_c^{x_2'}}{2\varepsilon\sqrt{k_e\beta_{nc}/2}} \left(\left(\frac{a_{2n}k_e\beta_{nn} - a_{2p}k_e\beta_{np}}{2} \right) \right.
$$

$$
\left. - \theta w \left(\frac{a_{1n}k_e\beta_{nn}(1 + \Delta_2')}{4(1 + k_{2n})} + \frac{a_{1p}k_e\beta_{np}(1 + \Delta_1')}{4(1 + k_{2p})} \right) \right)
\tag{6.34}
$$

4. The third-order harmonic, which is a source of intermodulation distortion, can be expressed as follows:

$$
H_{D3} \approx \frac{A^3 I_c^{x_2'}}{4\sqrt{(k_e\beta_{nc}/2)}} \left(\frac{-w(\Delta_{31}' - \Delta_{32}')}{4\varepsilon^2} \right)
$$

$$
\times \left(\frac{a_{2n}k_e\beta_{nn} - a_{2p}k_e\beta_{np}}{2} \right) + \frac{A^3 I_c^{x_2'}}{4\sqrt{k_e\beta_{nc}/2}} \left(\frac{w(\Delta_{31}' - \Delta_{32}')}{4\varepsilon^2} \right)
$$

$$
\times \left(\theta w a_{1n} \frac{k_e\beta_{nn}}{4(1 + k_{2n})}(1 + \Delta_2') + \theta w a_{1p} \frac{k_e\beta_{np}}{4(1 + k_{2p})}(1 + \Delta_1') \right)
$$

$$
+ \frac{\varepsilon^{-1} A^3 I_c^{x_2'}}{4\sqrt{k_e\beta_{nc}/2}} \left(\frac{a_{3n}k_e\beta_{nn} - a_{3p}k_e\beta_{np}}{2} \right) - \frac{\varepsilon^{-1} A^3 I_c^{x_2'}}{4\sqrt{k_e\beta_{nc}/2}}
$$

$$
\times \left(\frac{\theta w a_{2n}k_e\beta_{nn}}{4(1 + k_{2n})}(1 + \Delta_2') + \frac{\theta w a_{2p}k_e\beta_{np}}{4(1 + k_{2p})}(1 + \Delta_1') \right)
\tag{6.35}
$$

All these analytical results are illustrated from Fig. 6.8 to Fig. 6.10 using the design values of a filter design example given in Section 6.2.4. The total harmonic distortion (THD) due to the variations of the exponents (but in absence of mismatch) is shown in Fig. 6.8. The deviations in the exponent of the linear term in the NG-block will be the main source of distortion. This is smaller than 1 per cent and it can be

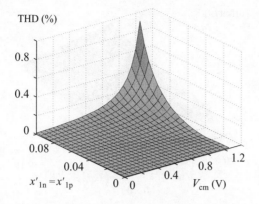

Figure 6.8   *Total Harmonic Distortion versus the common mode signal and the variation in the exponents of I_n and I_p (eq. (6.29), A=0.25 V)*

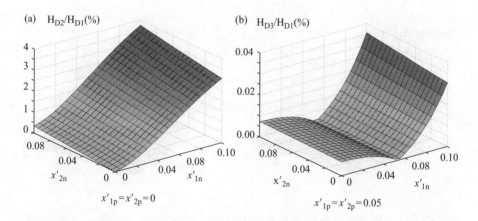

Figure 6.9   *Second- and third-order harmonics when the exponents x'_2 (eq. (6.29)) and x'_1 (eq. (6.29)) change in a different form in the negative and positive part of the NG-block*

as low as 0.2 per cent, provided that the common-mode signal is under 1V (which, on the other hand, has to be satisfied in order to prevent the output from saturation). Figure 6.9 to Fig. 6.11 show the different harmonics that result when not only one of the exponents differs from the ideal value, but also the deviation is different in the positive and the negative parts of the circuit (subscripts p and n, respectively). The second harmonic will dominate if mismatch exists between both parts. The figures show that the main source of distortion comes from the changes in the x'_1. Hence, the filter should be designed in such a way that the parasitics are minimised, especially in transistor M1, and also transistors M1 and M4 are well matched.

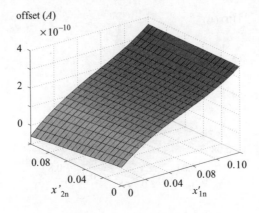

Figure 6.10   Offset versus different values of the exponents x'_2 and x'_1 for $G_m = 8e\text{-}10\ A/V$

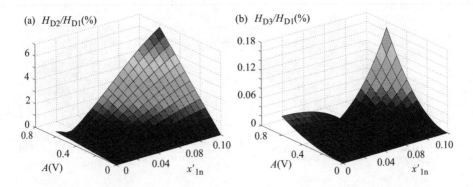

Figure 6.11   Second- and third-order harmonics versus input amplitude and deviation in the exponent of I_n when a 2 per cent mismatch is considered in the other parameters

The offset when $k_e = 1$ is shown in Fig. 6.10. Again, it is important to keep under control the mismatch between the capacitors that affect the exponents that set the state-space currents. Otherwise the offset could be a significant term. In any case, k_e will always be smaller than 1 and this will reduce its effect. All these figures have been obtained for an input amplitude $A = 0.25$, a common mode current of 225 nA and a common mode voltage of 0.625 V.

An example of the nonlinearities that arise when different input amplitudes are applied is shown in Fig. 6.11. Again, the matching in the exponents of I_n and I_p is the important factor to take into account. Hence, in general, the THD will be dominated by the second harmonic, even when the structure is fully differential. This term is mainly the consequence of the mismatch between the exponent of the single output currents when they are processed by the NG-block, so the capacitances involved in this ratio have to be designed carefully. Furthermore, the parasitics should be minimised, or at least they should be as equal as possible in both single sides.

6.2.3.2 PSRR+

This section analyses how the variations of the positive voltage supply can affect the performance of the integrator block. This study is important since V_{DD} is connected to some FGMOS inputs (these connections were required either to scale up the values of the currents (equation (6.22)) or to scale down the effective threshold voltage (equation (6.14)). Taking mismatch into account, eq. (6.6) gives rise to two new different equations for the positive and negative side:

$$\dot{V}_p - \dot{V}_n + \alpha_{11}V_p - \alpha_{12}V_n = \eta_{11}x_{ip} - \eta_{12}x_{in} \tag{6.36}$$

$$\dot{V}_p - \dot{V}_n + \alpha_{21}V_p - \alpha_{22}V_n = \eta_{21}x_{ip} - \eta_{22}x_{in} \tag{6.37}$$

Considering the case when the differential input signal is zero (this is valid as V_{DD} is taken as the input) the system of equations given by eq. (6.36) and eq. (6.37) can be solved for the differential output as a function of the common mode at the input to give[66]

$$V_d = \frac{(\alpha_{21} - \alpha_{22})(\eta_{11} - \eta_{12}) - (\alpha_{11} - \alpha_{12})(\eta_{21} - \eta_{22})}{(\alpha_{12}\alpha_{21} - \alpha_{11}\alpha_{22})} V_{icm} \tag{6.38}$$

This equation can be rewritten in a different form considering how the sources of mismatch are going to affect the different parameters. Hence, the difference between some of them is going to appear as just a multiplicative factor, whereas for others their difference is going to affect as an exponential function. Taking this into account, new variables can be defined in the following way:

$$\alpha_{21} = k_1'\alpha_{11}, \quad \alpha_{22} = k_2'\alpha_{12}, \quad \eta_{21} = k_3'\eta_{11}, \quad \eta_{22} = k_4'\eta_{12} \tag{6.39}$$

Also, the α and η parameters depend on V_{DD} as

$$\alpha_{1j} = \alpha_{1j}'e^{(w_{DDj}''/nU_T)V_{DD}}, \quad \eta_{1j} = \eta_{1j}'e^{(w_{DDj}'/nU_T)V_{DD}} \qquad \text{for } j = [1,2] \tag{6.40}$$

---

[66] The following derivations are only valid under the stated assumptions, and as long as all the devices in the blocks are operating in the right operating region. If the mismatch is such that the CMFB stops working or the cascode transistors in one of the branches leave the saturation region, these expressions would not be valid anymore. Also, usual contributions to the PSRR+ have been neglected and only those due to the FGMOS devices are considered. This is done for the sake of simplicity and also to give the reader an idea of how much the FGMOS devices on their own would contribute to the degradation of the performance.

$$V_d = \frac{(k_1'\alpha_{11} - k_2'\alpha_{12})(\eta_{11} - \eta_{12}) - (\alpha_{11} - \alpha_{12})(k_3'\eta_{11} - k_4'\eta_{12})}{\alpha_{12}\alpha_{11}(k_1' - k_2')}V_{icm} \quad (6.41)$$

Assuming that the variations of V_{DD} are much smaller than its nominal value eq. (6.41) can be approximated as

$$\frac{\Delta V_d}{\Delta V_{DD}}$$

$$\approx \frac{(k_1' - k_3')\alpha_{11}\eta_{11}(w_{DD1}' - w_{DD2}'') + (k_2' - k_4')\alpha_{12}\eta_{12}(w_{DD2}' - w_{DD1}'')}{nU_T\alpha_{11}\alpha_{12}(k_1' - k_2')}\bigg|_{V_{DD}} V_{icm}$$

$$+ \frac{(k_4' - k_1')\alpha_{12}\eta_{12}(w_{DD2}' - w_{DD2}'') + (k_3' - k_2')\alpha_{12}\eta_{11}(w_{DD1}' - w_{DD1}'')}{nU_T\alpha_{11}\alpha_{12}(k_1' - k_2')}\bigg|_{V_{DD}} V_{icm}$$

$$(6.42)$$

If the differential output/input gain is high enough in comparison with the percentage of mismatch and so the latter can be neglected in its expression, the PSRR+ is then given by

$$\text{PSRR+} = \frac{(nU_T/V_{icm})}{(w_{DD2}' - w_{DD2}'') + ((k_4' - k_3')/(k_1' - k_2'))(w_{DD1}' - w_{DD2}'')} \quad (6.43)$$

The main conclusion that can be extracted from this expression is that the PSRR+ can get seriously degraded because of the exponential relationship between the currents and the V_{DD} connected inputs in the square root block. Assuming the hypothetical case of the same percentage of mismatch of 1 per cent for all the devices, and taking that all the w_{DD} weights have the ideal value of 0.2 and $V_{icm} = 0.5$ V, the PSRR+ would be only 25 dB. A way to improve this value is to reduce the capacitive weight connected to V_{DD} or to connect that input to a low noise bias voltage instead.

6.2.4   A second-order filter example

This section describes the implementation of a second-order filter prototype for audio applications designed following the design procedure explained above. The power supply voltage was aimed to remain below 1.25 V, for a circuit implemented in a 0.8 μm CMOS technology with nominal values for the threshold voltages of $V_{Tn} = 0.8$ V and $|V_{Tp}| = 0.82$ V. The starting point for the design is equations of the form

$$\dot{x}_1 = -\omega_{o1}x_1 - \omega_{o2}x_2 + \omega_{o3}x_i \quad (6.44)$$

$$\dot{x}_2 = \omega_{o2}x_1 \quad (6.45)$$

In this case, three parameters, ω_{oi} for $i = [1, 3]$, determine the filter programmability. The LP and BP filter transfer-functions are as follows:

$$X_2(s) = \frac{\omega_{o3}\omega_{o2}}{s^2 + \omega_{o1}s + \omega_{o2}\omega_{o2}}X_i(s) \quad (6.46)$$

$$X_1(s) = \frac{s\omega_{o3}}{s^2 + \omega_{o1}s + \omega_{o2}\omega_{o2}}X_i(s) \quad (6.47)$$

which are related to the filters specifications as

$$\omega_o = \omega_{o2}, \quad Q = \frac{\omega_{o2}}{\omega_{o1}},$$

$$H_{LP}(0) = \frac{\omega_{o3}}{\omega_{o2}}, \quad H_{BP}(\omega_o) = \frac{\omega_{o3}}{\omega_{o1}} \quad (6.48)$$

A block diagram of a circuit that implements eqs. (6.44–6.45) is shown in Fig. 6.12, where

$$x_1 = (V_{1p} - V_{1n}) \quad (6.49)$$

$$x_2 = (V_{2p} - V_{2n}) \quad (6.50)$$

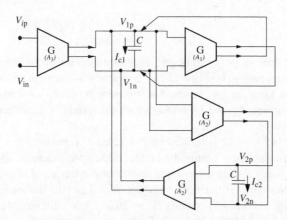

Figure 6.12   Second-order filter circuit with the proposed FGMOS transconductor

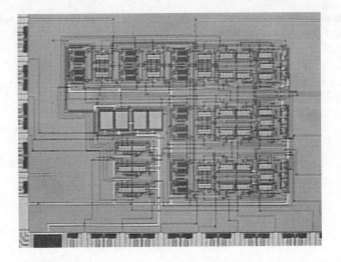

Figure 6.13   Photograph of the second-order filter

$$\omega_o = \frac{wA_2\sqrt{2\beta_n I_{A2}}}{C} \tag{6.51}$$

$$Q = \sqrt{\frac{I_{A2}}{I_{A1}}} \times \frac{A_2}{A_1} \tag{6.52}$$

$$H_{LP}(0) = \sqrt{\frac{I_{A3}}{I_{A2}}} \times \frac{A_3}{A_2} \tag{6.53}$$

$$H_{BP}(\omega_o) = \sqrt{\frac{I_{A3}}{I_{A1}}} \times \frac{A_3}{A_1} \tag{6.54}$$

Tuning of either ω_o or Q can be performed through either the reference voltages or the independent current sources. The ω_o frequency will remain constant if V_{A2} and I_{A2} are constant. Q can be programmed either through I_{A1} or V_{A1}. The circuit outputs are either the currents I_{c1} and I_{c2} through both capacitors or the differential voltage across them.

The circuit in Fig. 6.12 is formed by two equal C capacitors and four transconductors that are designed according to the filter specifications. Because of the low-frequency range chosen high output resistances are required in order to preserve the low time constants of the filter. A value of 3.4 pF for the integrating capacitances is chosen. For the NG-circuits $\beta_n = 30.7$ μA/V^2; for the M1–M3 FGMOS transistors in Fig. 6.3. 50 fF and 100 fF input capacitors implement the 0.25 and 0.5 weights, respectively. For the $y/\sqrt{x}$ circuits in Fig. 6.4 transistors were sized to give $\beta_1 = 972 \ \mu\text{A/V}^2$, $\beta_2 = \beta_3 = 322 \ \mu\text{A/V}^2$ and $\beta_4 = 41 \ \mu\text{A/V}^2$. These different aspects add a multiplicative factor to the τ coefficients, which has been omitted in the equations for the sake of simplicity. A photograph of the final circuit layout is shown in Fig. 6.13. The total area is $0.23 \ \text{mm}^2$.

Some experimental results that illustrate the performance of the filter are given below. Figure 6.14(a) and (b) represent the LP and BP transfer functions.

The distortion for the LP function can be observed in Fig. 6.15 for a sinewave input signal of 200 Hz in a filter with a cut-off frequency of 1 kHz. The maximum value of distortion is -41 dB for an input amplitude of $0.5 \ V_{\text{DD}}$.

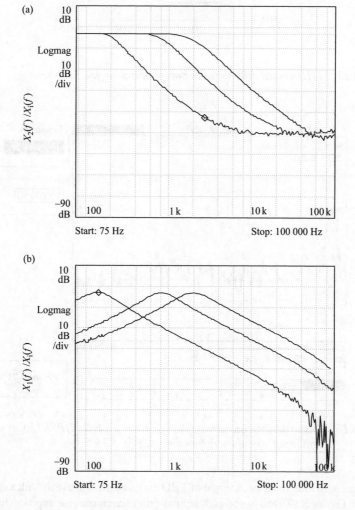

Figure 6.14   Experimental LP (a) and BP (b) filter

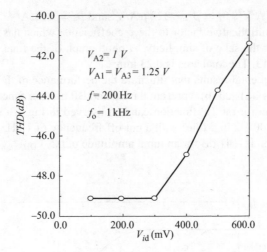

Figure 6.15   THD versus differential input amplitude in the LP filter

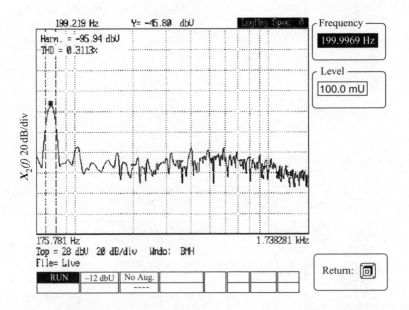

Figure 6.16   Distortion measured with the spectrum analyser HPSR770 for the filter and input signal corresponding to Fig. 6.15

Figure 6.16 illustrates an example of THD measurement obtained with a spectrum analyser. Figure 6.17 clearly shows a second-order harmonic that appears due to the mismatch between both symmetrical branches.

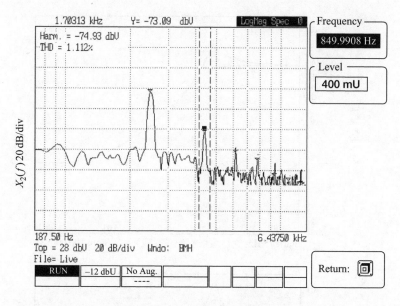

Figure 6.17 Second-order harmonic in the low-pass filter with Q = 1, gain = 1 and f₀ = 2 kHz

The performance of the filter is also verified when the input common mode changes. For this design the experimental CMR was 0.6 V–0.8 V. Figures 6.18 and 6.19 show the Q and gain programming for both transfer functions, respectively.

The noise floor is shown in Fig. 6.20. The cursor marks the output for a single input amplitude of 0.1 mV, which is used as the reference.

Results of the performance are summarised in Table 6.2. They are in agreement with the theoretical and simulated ones.

6.3 Summary and conclusions

The design procedure for a low power continuous-time linearised G_m-C filter based on the operation of the FGMOS transistor in the weak inversion region has been presented in this chapter. The technique is specially suitable for very low frequency applications in which dynamic range and power consumption are the most important issues. The advantages the FGMOS devices add to this kind of topology are as follows:

1. The operating point at the gate of the transistors can be shifted, this making possible their operation in the desired region for effective input levels that on its own would not meet the minimum requirements for that.

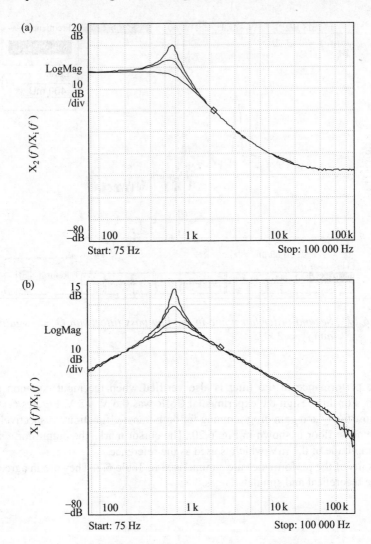

Figure 6.18   Q-programming for LP (a) and BP (b) functions for $f_0 = 0.75$ kHz

2. Translinear loops can be implemented without using the source terminal. Because of this only two transistors need to be stacked. This is very important when low voltage is a design constraint.
3. Less devices are needed to get a nonlinear ratio between currents, thanks to the different possible weights in the transistor.
4. The functions can be programmed by using different inputs in the transistors.
5. It facilitates the common mode sensing and feedback mechanism.

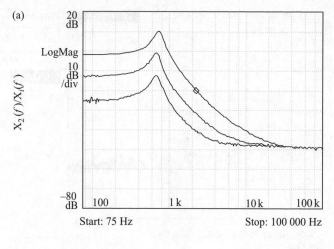

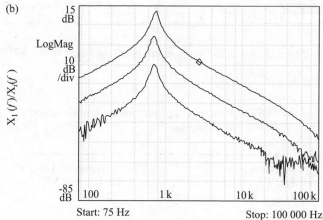

Figure 6.19   Gain-programming for LP (a) and BP (b) functions for $f_0 = 0.75$ kHz

However, these advantages do not come for free. There are two main disadvantages of using FGMOS instead of MOS for this type of circuit:

1. A larger area is needed because of the input capacitors.
2. Mismatch between them can cause undesired effects such as an increase in the PSRR+ as well as THD.

Therefore, it is ultimately up to the designer to decide what weighs more, the advantages or the disadvantages, in their specific design.

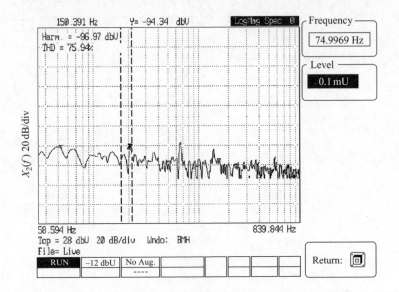

Figure 6.20   Noise floor. Cursor mark in an input amplitude of 0.1 mV

Table 6.2   Summary of the second-order filter
            performance

Technology	0.8 μm-AMS CXQ
V_{Tn}, V_{Tp}	0.8 V, −0.82 V
V_{DD}	1.25 V
Area	0.23 mm^2
$f_{o,min}$	100 Hz
$f_{o,max}$	2 kHz
Q_{min}	0.7
Q_{max}	7
THD$_{max}$($V_{PP} < 1$ V@200 Hz, $Q = 1, H_{LP}(0) = 1, f_o = 900$ Hz)	<-40 dB
IM3 ($V_{pp1} = V_{pp2} < 0.5$@200 Hz, $Q = 1, H_{LP}(0) = 1, f_o = 900$ Hz)	<-40 dB
Noise in band	−75 dB
DR	78–62 dB
PSRR+	>40 dB
Power	2.5 μW

Notation

α	Coefficient of the state variable in eq. (6.5)
α_{ij}	α parameter with mismatch (eqs. (6.36) and (6.37))
α'_{ij}	Terms in α_{ij} that do not depend on V_{DD}
β_{nc}	β_n parameter for transistor M3' in NG-block (Fig. 6.3)
β_{nn}	β_n parameter for transistor M2' in NG-block (Fig. 6.3)
$\beta'_{n(n,p)}$	See eq. (6.30)
β_{np}	β_n parameter for transistor M1' in NG-block (Fig. 6.3)
Δ'_{3i} for $i = [1, 2]$	Percentage of deviation with respect to its ideal value of the input capacitance connected to transistor Mi in transistor M3 (Fig. 6.3)
Δ_i	Mismatch between diode connected capacitances in transistor Mi (Fig. 6.4, eq. (6.28))
Δ'_i for $i = [1, 2]$	Mismatch between input capacitive weights, referring to transistor Mi' in NG-block (Fig. 6.3, eq. (6.25))
$\Delta V_{TFG(n,p)}$	See Table 6.1
ε	See Table 6.1
η	Coefficient of the input variable in the state-space equation (6.5)
η_{ij}	η parameter with mismatch (eqs. (6.36) and (6.37))
η'_{ij}	Terms in η_{ij} that do not depend on V_{DD}
τ	Time constant G_m/C
ω_0	Cut-off frequency (rad/s. Eq. (6.48))
ω_{oi} for $i = [1, 3]$	Coefficients in the state-space equations of the second-order filter prototype
A	Amplitude of a sinusoidal input signal
$A_{0(n,p)}$	See Table 6.1
$a_{o(n,p)}$	See Table 6.1
$A_{1(n,p)}$	See Table 6.1
$a_{1(n,p)}$	See Table 6.1
$A_{2(n,p)}$	See Table 6.1
$a_{2(n,p)}$	See Table 6.1
$A_{3(n,p)}$	See Table 6.1
$a_{3(n,p)}$	See Table 6.1
A_i	Non-dimensional gain (eq. (6.19)) related to the input term in the state-space equation
A_j	Non-dimensional gain (eq. (6.19)) related to the state variable term
C	Integrating capacitance
$A'_{j(n,p)}$	$A_{j(n,p)}$ parameter when there is mismatch that affects multiplicative constants in the expression of the currents (eq. (6.29) and Table 6.1)
$b_{1(n,p)}$	See Table 6.1

$B'_{(n,p)}$	See Table 6.1
c_i	$(\Delta_i + C_{GDi}/wC_{Ti})$. Contribution of mismatch as well as gate to drain parasitic capacitance to the effective input weight in transistor Mi (eq. (6.28))
$c_{i(n,p)}$	See Table 6.1
$C'_{(n,p)}$	See Table 6.1
$D'_{(n,p)}$	See Table 6.1
$E_{(n,p)}$	See Table 6.1
f_o	Cut-off frequency (Hz)
$F_{(n,p)}$	See Table 6.1
$G'_{(n,p)}$	See Table 6.1
$G_{m(NG)}$	Transconductance of NG-block. See Fig. 6.3
$H_{BP(\omega_o)}$	Bandpass filter gain at the cut-off frequency (eq. (6.48))
$H_{LP}(0)$	Lowpass filter DC gain (eq. (6.48))
I_0	Drain current in M0 (Fig. 6.1)
I_1	Drain current in M1 (Fig. 6.1)
I_{Ai}	Independent current in the input square root block (Fig. 6.4)
I_{Aj}	Independent current in the state-space square root block (Fig. 6.4)
I_c	Common mode current (eq. (6.7))
I_{c1}	Current through one of the integrating capacitors in Fig. 6.12
I_{c2}	Current through one of the integrating capacitors in Fig. 6.12
I_d	Differential current $I_p - I_n$
I_{ic}	Common mode input current (eq. (6.8))
I_{id}	Differential input current $I_{ip} - I_{in}$
I_{in}	Current related to V_{in} through eq. (6.8)
I_{ip}	Current related to V_{ip} through eq. (6.8)
I_n	Current related to V_n through eq. (6.7)
I_p	Current related to V_p through eq. (6.7)
I_{on}	Output current implementing the square root function for I_n (Fig. 6.4)
I_{op}	Output current implementing the square root function for I_p (Fig. 6.4)
K_1	Gain coefficient in the change of variables in eq. (6.7) and (6.8). $w\sqrt{2\beta_n}$
$k_{1(n,p)}$	See Table 6.1
k_2	$\theta[w(1 + \Delta'_{(3,4)})V_{cm} - V_{TFG(n,p)}]$
k_e	Scaling factor for the currents from the NG-block (from the gain of a current mirror)
k'_i for $i = [1,4]$	Parameters to account for mismatch in α_{ij} and η_{ij} that only affects as a multiplicative constant (eq. (6.39))
M	Transistor M1 number of inputs in Fig. 6.1
N	Transistor M0 number of inputs in Fig. 6.1

(n,p)	Subscript used to simultaneously refer to variables with subscripts n and p
Q	Quality factor (eq. (6.48))
V_{0i}	FGMOS inputs (transistor M0 in Fig. 6.1)
V_{1j}	FGMOS inputs (transistor M1 in Fig. 6.1)
V_{1n}	Single state variable (voltage) related to x_{1n} in eq. (6.49)
V_{1p}	Single state variable (voltage) related to x_{1p} in eq. (6.49)
V_{2n}	Single state variable (voltage) related to x_{2n} in eq. (6.50)
V_{2p}	Single state variable (voltage) related to x_{2p} in eq. (6.50)
V_{Aj}	Tuning voltage for the square root block (Fig. 6.4)
V_c	Constant voltage in one of the inputs of some FGMOS transistors in the square root block shown in Fig. 6.4
V_{cm}	Common mode voltage $(V_p + V_n)/2$ (Fig. 6.3)
v_{DD}	Variation of V_{DD}
V_d	$V_p - V_n$
V_i for $i = [1, N]$	Effective inputs in an FGMOS in weak inversion (eq. (6.2))
V_{id}	$V_{ip} - V_{in}$
V_{icm}	Common mode at the input
V_{in}	Single input in the state space-equation (voltage) (eq. (6.8))
V_{ip}	Single input in the state space-equation (voltage) (eq. (6.8))
V_n	Single state variable (voltage) (eq. (6.7))
V_{out}	Differential output voltage
V_{outcm}	Output of the CMFB
V_p	Single state variable (voltage) (eq. (6.7))
V_{TFG}	Effective threshold voltage of transistors in Fig. 6.3 (eq. (6.14))
V_{TFGc}	V_{TFG}. Effective threshold voltage with mismatch for transistor M'3 in the NG-block (Fig. 6.3). Taken as nominal value in the calculations (eq. (6.27))
$V_{TFG(n,p)}$	Effective threshold voltage with mismatch for transistors M'2 and M'1, respectively, in the NG-block (Fig. 6.3, eq. (6.27))
V_{Tn}	Threshold voltage for n-channel MOS devices (for the sake of simplicity it is assumed the same for all of them unless the opposite is said)
V_{Tp}	Threshold voltage for p-channel MOS devices (for the sake of simplicity it is assumed the same for all of them unless the opposite is said)
w	See w_{in}
w_{0k}	Effective input weight for the input k in transistor M0 (Fig. 6.1)
w_{1k}	Effective input weight for the input k in transistor M1 (Fig. 6.1)
w_{DD}	$w_{VDD} = C_{VDD}/C_T$. Weight of input connected to V_{DD} in transistors in the NG-block (Fig. 6.3)

$w_{DD(2,1)}$	Weights of inputs connected to V_{DD} in transistors M2$'$ and M1$'$ in NG-block when mismatch is considered (Fig. 6.3)
w''_{DDj} for $j = [1,2]$	Sum of all the weights that contribute to the V_{DD} term in α_{ij}
w'_{DDj} for $j = [1,2]$	Sum of all the weights that contribute to the V_{DD} term in n_{ij}
w_i	C_i/C_T. FGMOS input weights (eq. (6.2))
w_{in}	$C_{in}/C_T = w/2$. Weight of inputs connected to V_p and V_n in transistors in NG-block (Fig. 6.3)
W_{Ri}	$w(1 + c_i)$. Weight w when mismatch and gate to drain parasitic capacitance are taken into account in transistor Mi.
x	State variable in the state eq. (6.5)
x_i	Input variable in the state eq. (6.5)
x'_1	Parameter related to the percentage of mismatch as shown in eq. (6.29) and Table 6.1
x'_2	Parameter related to the percentage of mismatch as shown in eq. (6.29) and Table 6.1
x_j for $j = [1,2]$	State variables in eq. (6.44) and eq. (6.45)
x_{in}	Single input for the fully differential implementation in eq. (6.6)
x_{ip}	Single input for the fully differential implementation in eq. (6.6)
x_n	Single state variable for a fully differential implementation in eq. (6.6)
x_p	Single state variable for a fully differential implementation in eq. (6.6)

Chapter 7

Low power log-domain filtering based on the FGMOS transistor

The filtering operation performed by a linear, time invariant system (LTI) is mathematically described by a set of LTI equations as it has been shown several times in previous chapters. These equations relate two magnitudes: the input and the output of the system. However, they do not impose any mathematical constraint in internal variables, which implies that internal variables can be related in any manner and not necessarily in a linear way. This opens up a wider set of options in the internal processing of the signals. Among the nonlinear functions that can be chosen to process the internal variables, logarithmic and exponential functions have proven to be very interesting, mostly because they link directly to physical models of transistor devices.

The concept of log-domain signal processing was originally proposed by Adams [179] and rigorously formalised by Frey [182–184]. Frey proposed a nonlinear (exponential) mapping of the state variables in the state-space description of any linear transfer function that results in a set of nonlinear equations which can be interpreted according to the Kirchoff's current law. This class of circuits is known as Exponential State Space (ESS) circuits. Later on, Tsividis came up with a more generalised approach which included all the topologies that are Externally Linear Internally Nonlinear (ELIN) [21].

Log-domain filters can be found in many different applications due to the fact that an exponential function models the current of both a bipolar and a MOS transistor in the weak inversion region. The main difference between the two of them from the application point of view is that the current levels are different in both devices. Hence, BJTs can be used for high-frequency applications that require high current and transconductance values [189–193,223,224], whereas MOS are more suitable for low-frequency and low-power design [26,194,195,225]. Nevertheless, regardless of the application both of them offer an extended dynamic range under reduced power supply voltage as well as easy programmability [226].

Previous chapters have shown how the FGMOS in weak inversion also behave in an exponential way, but simultaneously many more parameters control the characteristics of the exponential function. Besides, as it was described in Chapter 3 a much more complex nonlinear processing than a simply exponential function can be performed within a single FGMOS device. This can be used advantageously on the implementation of log domain, low power, low frequency circuits. This chapter will explain how to design a log-domain filter using FGMOS transistors. It starts with a review of the basic log-domain principle, followed by the description of basic FGMOS blocks that implement the required functions. The blocks are afterwards put together on a basic lossy integrator circuit which serves as the basis of a higher order filter. As usual, the advantages and disadvantages of using FGMOS devices in these kinds of topologies will be discussed along the text. Also, as these topologies are internally nonlinear any deviation with respect to the ideal behaviour in the transistors, caused by for example mismatch, will give rise to strongly nonlinear terms that can degrade the performance in terms of distortion. A qualitative and quantitative analysis of second-order effects focused on the latter will follow. The chapter ends with a real design example that illustrates the performance of a log-domain second-order filter designed following the previously developed theory.

7.1   Log-domain integration with FGMOS

7.1.1   Basic principle

Let us start again with the general equation of a linear integrator:

$$\dot{x} = \alpha x + \eta x_{in} \tag{7.1}$$

In the general formulation of logarithmic filters [182], both the state variable, x, and the input, x_{in}, are mapped into two new variables, y and u, through the exponential relationships:

$$x = k_y e^{y}$$
$$x_{in} = k_{in} e^{u} \tag{7.2}$$

Hence eq. (7.1) can be rewritten as a nonlinear function of u and y:

$$\dot{y} = \alpha + K e^{(u-y)} \tag{7.3}$$

$$K = \frac{\eta k_{in}}{k_y} \tag{7.4}$$

If the terms in eq. (7.3) are associated to real physical variables, the left-hand side could represent the current flowing through a capacitor, in which case y would be proportional to the voltage difference between its terminals. The state variable, x, could be the current flowing through a device with an 'exponential behaviour', as for example a BJT transistor biased in the active region or an MOS transistor in the weak inversion saturation region (these two would be the traditional choices). Another

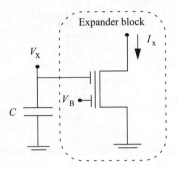

Figure 7.1   State variable definition: expander block

option is to use the FGMOS transistor also biased in the weak inversion saturation region, as it is shown in the following section.

7.1.2 Basic FGMOS circuits

This section describes basic circuit blocks that realise the mathematical transformations described before.

Figure 7.1 shows the schematic of a circuit block that generates the state variable, which in this case is the current I_x. The latter is the current in the channel of an FGMOS transistor with one of its inputs connected to a capacitor. A second input is connected to a constant voltage V_B. The mathematical function describing this current, assuming that the transistor is operating in the weak inversion saturation region and also that the ratio between the parasitic capacitances and the total capacitance observed by the FG is negligible[67], is

$$I_x = I_s e^{w_B V_B / n U_T} e^{w_x V_x / n U_T} = k_y e^{w_x V_x / n U_T} = k_y e^y \tag{7.5}$$

where w_B and w_x are the weights for the corresponding inputs. Hence, y is proportional to the voltage across the capacitor C and k_y depends exponentially on the voltage V_B. As in other logarithmic implementations the current 'expands' the input voltage because of the exponential function relating the two of them. Therefore, this transistor will be called expander from now on.

The input variable will also be a current; therefore, it will have to generate a voltage that is related to it in a logarithmic way (as shown in eq. (7.2)). Consequently if the argument of the logarithmic function is larger than 1 then big variations in this current will translate into smaller variations of the voltage. This voltage can be generated by an FGMOS transistor with one of its inputs connected to its drain. Because of the aforementioned 'compression' function, this device will be referred to as compressor.

---

[67] This will be assumed from now on unless the opposite is said.

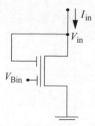

Figure 7.2   Generation of the input voltage from the input current

The circuit block is represented in Fig. 7.2. The current I_{in} can be written as a function of the voltages, yielding[68]

$$I_{in} = I_s e^{w_{Bin}V_{Bin}/nU_T} e^{w_{in}V_{in}/nU_T} = k_{in} e^{w_{in}V_{in}/nU_T} = k_{in} e^u \tag{7.6}$$

And V_{in} will be related to the input current in a logarithmic way as it was explained before:

$$V_{in} = \frac{nU_T}{w_{in}} \ln\left(\frac{I_{in}}{I_s}\right) - \frac{w_{Bin}}{w_{in}} V_{Bin} \tag{7.7}$$

where w_{in} and w_{Bin} are the weights corresponding to the inputs V_{in} and V_{Bin}, respectively.

Equation (7.3) can now be rewritten considering the already defined physical variables as shown in eq. (7.8). The RHS is the sum of two terms, a constant current plus a term that is proportional to the ratio between the input current and the state current. These currents are added into the integrating capacitance:

$$\frac{w_x \dot{V}_x}{nU_T} = \alpha + \eta \frac{I_{in}}{I_x} \tag{7.8}$$

Therefore, the following step would be to generate the ratio between the input and the state current, which can be done with the help of the translinear principle explained in the previous chapter. The schematic of the divider block is shown in Fig. 7.3. It works as follows: both transistors are equally sized and have the same value input capacitances. Assuming that all the previously defined weights are also the same, w, the output current I_{out1} is

$$I_{out1} = I_A e^{w(V_{in}-V_x)/nU_T} = I_A \frac{I_{in}}{I_x} e^{w(V_B-V_{Bin})/nU_T} \tag{7.9}$$

which is proportional to the ratio between the input and the state-space current.

---

[68] Both transistors have been assumed to be of the same size. This will be assumed from now on for all the devices unless the opposite is said.

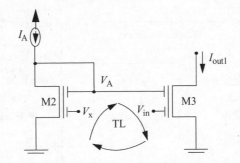

Figure 7.3   *Nonlinear block generating the ratio between the currents of the compressor and expander (NLB)*

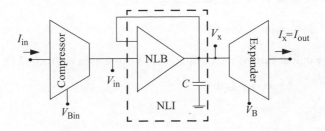

Figure 7.4   *Logarithmic integrator block diagram*

7.1.3   The integrator

All the previous blocks can be put together to form the lossy integrator described by eq. (7.1). A block diagram that illustrates the interconnections between them is shown in Fig. 7.4. The input block is the compressor. This circuit takes the input current and converts it logarithmically to the voltage V_{in}. The second block performs a nonlinear integration (NLI) as in (7.8). This block is formed by the nonlinear block (NLB), whose output is the ratio between the state and input currents, and the integrating capacitor C. The output signal is finally obtained taking the voltage at the integration node and processing it through the expander. The schematic for the final circuit is shown in Fig. 7.5.

In reality, this block is implemented by sourcing instead of sinking current to the integrating capacitance (for the sake of simplicity). This is done just by using a conventional current mirror to copy and invert the sign of the current delivered by M3 and connecting its output together with an inverted version of I_B to the integrating capacitance.

The expressions of the constants α and η can be obtained from Fig. 7.5 just by writing the equation of the voltage V_x in the integrating capacitance, rearranging it

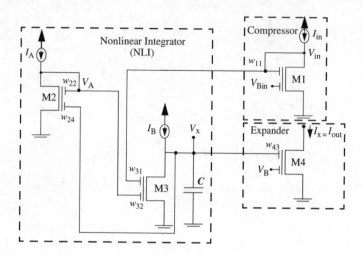

Figure 7.5   Integrator schematic[69]

to obtain the output current and comparing this expression with eq. (7.1):

$$\dot{V}_x = \frac{I_B}{C} - \frac{I_A}{C} e^{w(V_B - V_{Bin})/nU_T} \times \frac{I_{in}}{I_x} \tag{7.10}$$

$$\dot{I}_x = \underbrace{\frac{w}{nU_T} \times \frac{I_B}{C}}_{\alpha} I_x - \underbrace{\frac{w}{nU_T} \times \frac{1_A}{C} e^{w(V_B - V_{Bin})/nU_T} I_{in}}_{\eta} \tag{7.11}$$

In order to improve the dynamic range of the integrator, a differential version of it might be required. The differential version can be obtained taking as starting point a differential form of the state-space equations

$$(\dot{x}_p - \dot{x}_n) = \alpha(x_p - x_n) + \eta(x_{inp} - x_{inn}) \tag{7.12}$$

where the subscripts p and n have been used to differentiate between the positive and negative parts of the differential signal. For a lossless integrator, eq. (7.12) becomes

$$(\dot{x}_p - \dot{x}_n) = \eta(x_{inp} - x_{inn}) \tag{7.13}$$

In reality eq. (7.13) can be obtained by subtracting two single equations, one for the positive and another one for the negative side, respectively. Each one of the single equations can be realised using the procedure explained before. The subtraction can be performed just by inverting the currents with current mirrors. A schematic of a circuit that realises eq. (7.13) is shown in Fig. 7.6(a). Its block diagram is in Fig. 7.6(b). The + and − signs in the NLB have been used to differentiate between positive and negative currents. The differential version of the compressor is formed by transistors $M1_p$ and $M1_n$. The expander is constituted by transistors $M4_p$ and $M4_n$.

---

[69] Parameters w_{ij} are related to mismatch and are explained in a further section.

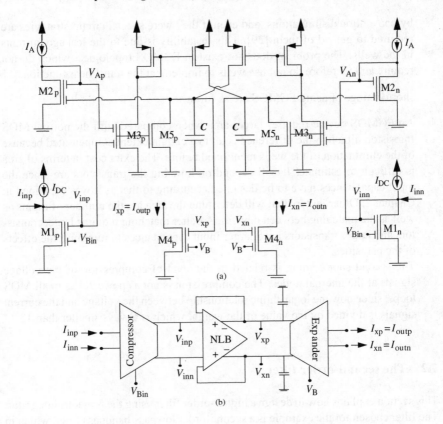

Figure 7.6   (a) Schematic of the fully differential integrator. (b) Block diagram

7.1.4   Advantages and disadvantages of using FGMOS transistors

The schematic of the integrator in Fig. 7.6 outlines some of the advantages of using FGMOS transistors:

1. First of all, neither the source nor the bulk terminal is required to implement the translinear equation. Because of this, it is possible to fix them to ground (or the negative power supply) as well. Therefore, there is no need for stacking transistors to implement the translinear loop, which is important when dealing with tight voltage and power constraints.

2. Besides, as the source and bulk terminals are connected to constant voltage values, V_{SB} turns out to be constant as well (in this case equal to zero) and separate wells do not have to be used anymore to reduce distortion[70]. It has been shown that sometimes these wells can cause oscillations in the output that are not predictable

---

[70] The gate to source voltage as a function of the current is also a function of V_{SB}, which causes distortion unless these terms are cancelled when implementing the translinear loop.

by conventional simulations, and even if they were special circuit strategies are required to get rid of them [29]. This instability is due to the leakage currents in the wells. The problem does not exist in FGMOS topologies, which do not require and therefore do not use wells to implement the translinear function.

Shortcomings of using FGMOS are as follows:

1. An FGMOS device requires larger area when compared with the normal MOS transistor, although the difference is up to a certain extent compensated because of the elimination of the wells mentioned before. The extra cost in terms of area is difficult to estimate. It will depend on how big the transistors are since the input capacitances have to be designed according to that as it was explained in Chapter 2. Other factors that will determine the area are the number of inputs as well as the matching between devices. A better matching requires larger transistors and larger transistors need larger input capacitances to minimise the effects of the parasitics.

2. The second shortcoming is related to the level of compression of the voltage signals at the internal nodes. The compression is not as powerful as in all MOS loops, since now the logarithmic relationship between the voltage and the current signals is divided by the value of the weight which is always smaller than 1.

7.2 The second-order filter

This section explains how to design a higher-order filter using the previous integrator. The filter chosen for the example is a second-order lowpass/bandpass filter, which in the time domain is described by the state-space eqs. (7.14) and (7.15):

$$(\dot{I}_{1p} - \dot{I}_{1n}) = -\omega_{o1}(I_{1p} - I_{1n}) - \omega_{o2}(I_{3p} - I_{3n}) + \omega_{o3}(I_{inp} - I_{inn}) \qquad (7.14)$$

$$(\dot{I}_{3p} - \dot{I}_{3n}) = \omega_{o4}(I_{1p} - I_{1n}) \qquad (7.15)$$

Equations (7.14) and (7.15) can be transformed applying changes of variables as in eq. (7.2). After doing so, the resulting expressions are[71] as follows:

$$\frac{w}{nU_T}(\dot{V}_{1p} - \dot{V}_{1n}) = -\omega_{o1} - \omega_{o2}(I_{3p}I_{1p}^{-1} - I_{3n}I_{1n}^{-1}) + \omega_{o3}(I_{inp}I_{1p}^{-1} - I_{inn}I_{1n}^{-1})$$

$$(7.16)$$

$$\frac{w}{nU_T}(\dot{V}_{3p} - \dot{V}_{3n}) = \omega_{o4}(I_{1p}I_{3p}^{-1} - I_{1n}I_{3n}^{-1}) \qquad (7.17)$$

A physical realisation of eqs (7.16) and (7.17) is shown in Fig. 7.7. This schematic has been a bit simplified for the sake of clarity. Hence, for example the values of ω_{oi}

---

[71] The changes of variables have been carried out in the single equations first and subsequently these single equations have been subtracted.

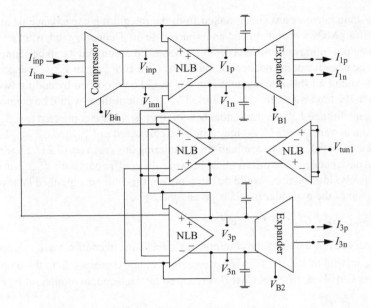

Figure 7.7   Block diagram of the second-order filter

have been designed to be the same for $i = [2, 3]$:

$$\omega_{o1} = \frac{w}{nU_T C} \times I_A$$

$$\omega_{oi} = \frac{w}{nU_T C} \times I_A e^{w(V_{B1} - V_{Bin})/nU_T} \quad i = [2, 3] \tag{7.18}$$

$$\omega_{o4} = \frac{w}{nU_T C} \times I_A e^{w(V_{B2} - V_{Bin})/nU_T}$$

The way to make ω_{o2} different from ω_{o3} is either to change the current I_A for each NLB, or as it is done in the design example at the end of the chapter, to add an extra input to the output transistor in the NLB (M3 in Fig. 7.6) and connect it to a tuning voltage V_{tuni}. If this is the choice, but still all the FGMOS transistors are wanted to be identical for the sake of matching, the extra input can be added to all of them. It can be connected to a constant voltage V_{cm} in those devices in which this input is not required. Under these circumstances, the general output of an NLB with an applied tuning voltage V_{tuni} is

$$I_{out1i} = I_A e^{w(V_{in} - V_x)/nU_T} e^{w(V_{tuni} - V_{cm})/nU_T}$$

$$= I_A \frac{I_{in}}{I_x} e^{w(V_B - V_{Bin})/nU_T} e^{w(V_{tuni} - V_{cm})/nU_T} \tag{7.19}$$

Also, in the final design another input can be added in order to compensate for common mode variations. This input will be connected to a voltage generated by a CMFB. The output of the CMFB only has to be connected to those FGMOS transistors whose drain currents are either sinked from the integrating capacitance or mirrored through a pMOS current mirror and sourced to it. Depending on whether it is a transistor providing current that is going to be sinked or sourced it will be connected to either the negative or the positive output of the CMFB, $V_{\text{outcm}}^{\text{p}}$ or $V_{\text{outcm}}^{\text{n}}$, respectively. Hence, taking all this into account Fig. 7.6 would be modified by adding two extra inputs to the transistors. In M1$_{(\text{n,p})}$ and M2$_{(\text{n,p})}$, both of them would be connected to a constant voltage V_{cm}. In transistors M5$_{(\text{n,p})}$ they would be connected to $V_{\text{outcm}}^{\text{p}}$ and V_{tuni}, and in transistors M3$_{(\text{n,p})}$ they would be connected to $V_{\text{outcm}}^{\text{n}}$ and V_{tuni}. The new coefficients of the state-space equations considering this are in eq. (7.21). Superscripts p and n have been used to differentiate between the coefficients with $V_{\text{outcm}}^{\text{p}}$ or $V_{\text{outcm}}^{\text{n}}$, respectively. Ideally, they would be the same but this will be explained later on. G_{ex} is the gain of the expander block as given by eq. (7.20):

$$I_{\text{out}(n,p)} = I_{\text{s}}G_{\text{ex}}e^{3wV_{\text{cm}}/nU_{\text{T}}}e^{wV_{x(n,p)}/nU_{\text{T}}} \tag{7.20}$$

The term corresponding to the common mode feedback in coefficient ω_{o4} depends on $V'^{(n,p)}_{\text{outcm}}$ instead of $V^{(n,p)}_{\text{outcm}}$. This is because this voltage corresponds to the output of a different CMFB (a different CMFB is required for each couple of integrating nodes):

$$\omega_{o1} = \frac{wG_{\text{ex}}}{nU_{\text{T}}C}I_{\text{A}}e^{wV_{\text{tun1}}/nU_{\text{T}}}e^{-2wV_{\text{cm}}/nU_{\text{T}}}e^{wV^{(n,p)}_{\text{outcm}}/nU_{\text{T}}}$$

$$\omega_{o2} = \frac{wG_{\text{ex}}}{nU_{\text{T}}C}I_{\text{A}}e^{wV_{\text{tun2}}/nU_{\text{T}}}e^{-2wV_{\text{cm}}/nU_{\text{T}}}e^{wV^{(n,p)}_{\text{outcm}}/nU_{\text{T}}}$$

$$\omega_{o3} = \frac{wG_{\text{ex}}}{nU_{\text{T}}C}I_{\text{A}}e^{wV_{\text{tun3}}/nU_{\text{T}}}e^{-2wV_{\text{cm}}/nU_{\text{T}}}e^{wV^{(n,p)}_{\text{outcm}}/nU_{\text{T}}} \tag{7.21}$$

$$\omega_{o4} = \frac{wG_{\text{ex}}}{nU_{\text{T}}C}I_{\text{A}}e^{wV_{\text{tun4}}/nU_{\text{T}}}e^{-2wV_{\text{cm}}/nU_{\text{T}}}e^{wV'^{(n,p)}_{\text{outcm}}/nU_{\text{T}}}$$

A further modification to the filter in Fig. 7.7 aimed to reduce power consists of eliminating transistor M2 in the NLB realising the coefficient ω_{o1} and connecting the two inputs in transistor M3 to V_{tun1}. The new ω_{o1} coefficient is then given by

$$\omega_{o1} = \frac{w}{nU_{\text{T}}C}I_{\text{s}}e^{3wV_{\text{tun1}}/nU_{\text{t}}}e^{wV^{(n,p)}_{\text{outcm}}/nU_{\text{T}}} \tag{7.22}$$

7.2.1   The expander

The expander block for the filter in Fig. 7.7 is a modification of the simple basic expander circuit in Fig. 7.1. The schematic for this circuit is shown in Fig. 7.8. One of the inputs in each FGMOS is connected to either a positive or a negative integrating node. The other three inputs are connected to a constant voltage, V_{e}, which can also be used to adjust the signals levels. The output current is hence given by

$$I_{\text{outj}} = I_{\text{jp}} - I_{\text{jn}} = I_{\text{s}}e^{3wV_{\text{e}}/nU_{\text{T}}}(e^{wV_{\text{jp}}/nU_{\text{T}}} - e^{wV_{\text{jn}}/nU_{\text{T}}})$$

$$= I_{\text{s}}G_{\text{ex}}e^{3wV_{\text{cm}}/nU_{\text{T}}}(e^{wV_{\text{jp}}/nU_{\text{T}}} - e^{wV_{\text{jn}}/nU_{\text{T}}}) \quad j = 1,3 \tag{7.23}$$

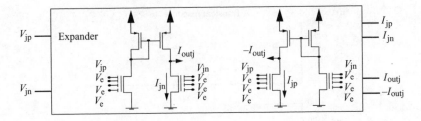

Figure 7.8   Fully differential expander

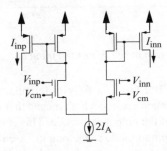

Figure 7.9   Input stage

7.2.2   The input stage

Although all the previously described blocks are operating in the current mode sometimes the input signal might be a voltage and a voltage to current conversion would be required. A simple circuit that can be used with this aim and proved to work with this filter (obviously under certain limitations) is shown in Fig. 7.9. Despite the transistors being operating in the weak inversion region, which makes the differential pair strongly nonlinear, the compression of the input voltage carried out at the FG by the ratio between the effective input capacitance and the total capacitance enables these to have relatively large signals at the input. However, it should be kept in mind that the dynamic range remains the same as the equivalent input noise is amplified by the inverse of the input weight.

An example of the performance of this block designed in a 0.35 μm technology is shown in Fig. 7.10. The input transistors aspect ratio is 10 μm/10 μm and the input capacitances have a value of 120 fF [217]. The effective input weight is 0.25. Fig. 7.10(a) illustrates the compression at the FG. Fig. 7.10(b) shows the output current. The THD is lower than 0.3 per cent for maximum signal levels of 200 mV.

7.3   The common mode control

It is a well-known fact that in order to avoid degradation in the frequency response of a G_m-C filter the output resistance in parallel with the integrating capacitance has

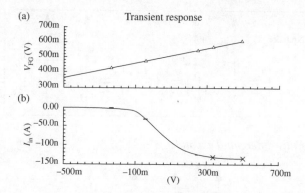

Figure 7.10   (a) Compression at the FG in the input stage. (b) Output current

to remain high enough[72]. But this is a fully differential structure with transistors operating in the weak inversion region, which means that mismatch could seriously affect the expected value of the output resistance. This is one of the reasons to have a circuit to control the common mode at the output. A circuit that works well with this filter is shown in Fig. 7.11. It is a p version of the CMFB described in Chapter 5. The principle of operation is the same. The only two differences between the two of them is that the bias current might be lower in this case and the transistors will then operate in the weak inversion region and also both differential outputs V_{outcm}^p and V_{outcm}^n are required in the feedback mechanism. These outputs are connected to transistors that supply current to the integrating capacitance. Depending on whether it is sinked or sourced current the connection should be to either one or the other. Any difference between the common mode level and V_{cm} creates a difference between the output voltages. This tends to generate different values for the currents provided to the integrating capacitances by the n and p branches. The difference has to be corrected in response to the Kirchoff's current law. The output resistance controls the change in the voltage which moves the operating point towards the reference voltage.

7.4   Design considerations. Second-order effects

7.4.1   Multiple operating points in the filter

Having separate wells in log-domain CMOS filters can provoke instabilities and make the output go into oscillation. This sometimes cannot be predicted by conventional simulations during the design process. Fox detected the problem while testing some circuits in the lab and proposed a method to detect this behaviour [203]. Log-domain filters implemented with FGMOS do not need any extra well to realise the translinear

---

[72] How high will depend on the characteristics of the filter.

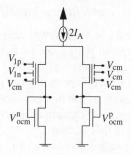

Figure 7.11   Common Mode Feedback Block (CMFB)

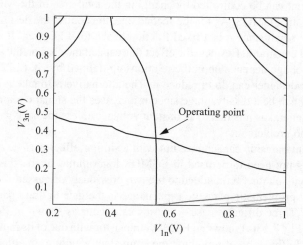

Figure 7.12   Analysis of stability in the filter

equation, and therefore do not have this problem. This is illustrated in Fig. 7.12 for the filter in Fig. 7.7 using Fox's method. The integrating capacitances are substituted by voltage sources which are swept in the range of interest. Subsequently, the currents flowing through these sources are printed and processed with the CONTOUR function in Matlab to find the crossing points in the integrating nodes when the currents are equal to zero. The latter would determine the possible operating points. If more than one exists (as it happens in normal MOS logarithmic filters) their stability would have to be analysed. The graph in Fig. 7.12 shows a single operating point, thus proving that the filter is inherently stable[73].

---

[73] Although the graph represents only two of the single outputs the results for the others are equivalent, showing also a single crossing point.

7.4.2 *Effect of the parasitic capacitances: qualitative study*

As it was explained in Chapter 2 one of the drawbacks of using FGMOS is the larger area of the device when compared with a conventional MOS transistor, owing to the input capacitors. The extra percentage of the total area can be made as small as the topology and performance tradeoffs permit, since parameters such as THD, gain and output resistance will degrade as the input capacitors are made smaller. This is because of the effect of the parasitic capacitances on the circuit performance. The latter is qualitatively analysed in this section.

Let us start with the gate to bulk capacitance C_{GB}. C_{GB} has two different components, one of them depending on the transistor size (this is the same as in a normal MOS transistor) and the other one due to the coupling between the poly1 bottom plate (FG) and the bulk (or well, in case the outer part of the FG has a well underneath). The first component can be controlled by changing the total area of the MOS transistor used to build the FGMOS. As for the second one, the control the designer has on its value is quite small since it is a fixed fraction of the total FG area. It is therefore a technological constraint. Besides, the effect this capacitance has on the scaling factor of the gate voltage depends upon the relationship defined by eq. (2.22), so there is not much the designer can do to reduce it. The alternative is to take it into account. Sometimes it can be a disadvantage since it increases the signal compression at the gate but, in general as it is tied to a constant voltage (unless the back gate is used), it will not be too problematic.

A different parasitic capacitance but with a similar effect is C_{GS}. As the source terminal does not need to be used in FGMOS logarithmic filters, this capacitance contributes with another term added to the two previously discussed.

The most problematic parasitic capacitance is the gate to drain parasitic capacitance, C_{GD}. Three different case scenarios can occur as shown in Fig. 7.13. The schematic in Fig. 7.13(a) shows an FGMOS transistor with one of its inputs connected to the drain. In this kind of topology, the equivalent weight to the diode connected input is not C_{in}/C_T, as it would be ideally assumed, but $(C_{in} + C_{GD})/C_T$. Since the translinear loop is designed to perform a nonlinear relationship between currents, and the exponents in this function are obtained with ratios between capacitances, a change either in the numerator or in the denominator of this ratio will alter the values of the exponents and hence the function. This can be a source of distortion and will be studied later on.

Figure 7.13(b) shows an FGMOS with a pMOS current mirror as a load. The output resistance has now an additional component in parallel equal to $(g_m C_{GD}/C_T)^{-1}$. Keeping this value much smaller than the pMOS transistor transconductance its effect will be negligible.

Figure 7.13(c) shows an NLB with an integrating capacitance at the output. If C_{GD} is assumed to be the same for all the FGMOS transistors, the voltage, V_A, is given by

$$V_A = \frac{nU_T}{w_D + w} \ln\left(\frac{I_A}{I_s}\right) - \frac{w}{w_D + w}V_x \tag{7.24}$$

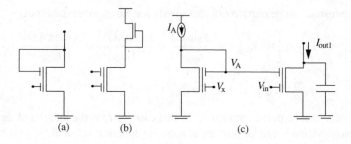

Figure 7.13 *FGMOS topological arrangements in the log-domain integrator. Analysis of the output resistance effects*

where $w_D = C_{GD}/C_T$. Hence, the current I_{out1} is

$$I_{out1} = I_s \left(\frac{I_A}{I_s} \right)^{w/(w+w_D)} e^{wV_{in}/nU_T} \cdot e^{-w^2 V_x/(w+w_D)} \tag{7.25}$$

Equation (7.25) shows how both the gain of the function as well as the value of the exponent are related to I_x change. This is going to translate on a variation of the integrator time constant as well as distortion. Besides, the time constant will depend on the technological parameter I_s. The effects of C_{GD} can be reduced either by increasing the minimum input capacitance or by limiting the swing at the drain node by using, for example, cascode transistors. The drawback of the latter is that stacking transistors makes the minimum voltage supply constraint more demanding.

7.4.3 Effect of C_{GD} and the mismatch between the inputs: a quantitative study

The effects of mismatch between the input capacitances and C_{GD} on the overall transfer function can be analysed using the same general functional dependence. This is because qualitatively they affect the performance of the block in the same way; although, quantitatively they can be very different. What will happen in general (either because of C_{GD} or mismatch or both) is that the relative weights will change with respect to their nominal values. This means that instead of getting the linear ratio between the input current and the state variable current shown in eq. (7.10), the ratio will be between non-integer powers of them. Taking this into account eq. (7.11) can be rewritten as

$$\dot{I}_x = \alpha' I_x - \eta' (I_{in})^{1+x_1} \cdot I_x^{-x_2} \tag{7.26}$$

where α' and η' represent the modified expressions of α and η and if mismatch and C_{GD} are taken into account. Their exact equations are irrelevant at the moment, so they will not be given for the sake of clarity.

As $x_1 \ll 1$, a Taylor series expansion can be performed. Also if an input consisting of a DC ($I_{in_{DC}}$) plus a sinusoidal term ($A \sin \omega t$) is considered and the amplitude of the varying term is as a first approximation much smaller than the DC term a second series expansion can be carried out that yields for the exponential input:

$$(I_{in})^{x_1} \approx 1 + x_1 \left[(\ln I_{inDc}) + \frac{A \sin \omega t}{I_{inDC}} - \frac{(A \sin \omega t)^2}{2(I_{inDC})^2} + \frac{(A \sin \omega t)^3}{3(I_{inDC})^3} + \cdots \right]$$

$$+ \frac{x_1^2}{2} (\ln I_{inDC})^2 + \frac{x_1^3}{6} (\ln I_{inDC})^3 + \cdots \qquad (7.27)$$

The same can be performed for I_x. In this case, as I_x is the integrated signal, as a first approximation it can be written as a cosine function ($I_x = I_{xDC} + A_1 \cos(\omega t)$). Then, according to eq. (7.6) the integrator equivalent input is now given by

$$(I_{in})^{1+x_1} \cdot (I_x)^{-x_2} \qquad (7.28)$$

It consists of different harmonics (apart from the fundamental one) which will generate distortion. The expression for the resulting THD considering as dominant the third-order harmonic is given by the following equation:

$$\text{THD} \approx 100 \sqrt{ \begin{aligned} & [(b_0 b_5 b_6 + b_1 b_4 b_6 - b_1 b_2 b_8)^2 + (b_0 b_9 b_2 - b_4 b_0 b_7 - b_1 b_3 b_7)^2] \\ & \times [16[(b_0 b_3 b_6 + b_1 b_2 b_6 + \frac{3}{4}(b_0 b_5 b_6 + b_1 b_4 b_6 - b_1 b_2 b_8) \\ & + b_1 b_2 b_8)^2 + (b_0 b_2 b_7 + b_4 b_0 b_7 \\ & + \frac{3}{4}(b_0 b_9 b_2 - b_4 b_0 b_7 - b_1 b_3 b_7) + b_1 b_3 b_7)^2]]^{-1} \end{aligned} }$$

$$(7.29)$$

The values of the parameters are shown in Table 7.1.

Let us now analyse briefly which are the factors that could make x_1 and x_2 different from zero. The names used for the different weights are shown in Fig. 7.5. Due to the

Table 7.1　*Equations for the different parameters used in the THD calculations*

b_0	I_{inDC}
b_1	A
b_2	$1 + x_1(\ln I_{inDC}) + \frac{x_1^2}{2}(\ln I_{inDC})^2 + \frac{x_1^3}{6}(\ln I_{inDC})^3 + \cdots$
b_3	$(x_1 A)/I_{inDC}$
b_4	$-x_1 A^2 / 2I_{inDC}^2$
b_5	$(x_1 A^3)/3I_{inDC}^3$
b_6	$1 - x_2(\ln I_{1DC}) + \frac{x_2^2}{2}(\ln I_{1DC})^2 - \frac{x_2^3}{6}(\ln I_{1DC})^3 + \cdots$
b_7	$-x_2 A_1 / I_{1DC}$
b_8	$x_2 A_1^2 / (2I_{1DC}^2)$
b_9	$-x_2 A_1^3 / (3I_{1DC}^3)$

gate to drain capacitance in transistor M1($C_{\text{GD}}^{(M1)}$, where the superscript refers to the name of the transistor) the input weight is

$$w_{11} = \frac{C_{\text{in}} + C_{\text{GD}}^{(M1)}}{C_T} = \frac{C_{\text{in}}}{C_T}\left(1 + \frac{C_{\text{GD}}^{(M1)}}{C_{\text{in}}}\right) \tag{7.30}$$

Also, if the input capacitance is connected to V_{in} in M3, it differs from the ideal value C_{in}/C_T as:

$$w_{31} = \frac{C_{\text{in}}}{C_T}(1 + \Delta_1) \tag{7.31}$$

Then the exponent of the input current is

$$(1 + \Delta_1)\Big/\left(1 + \frac{C_{\text{GD}}^{M1}}{C_{\text{in}}}\right) = 1 + x_1 \tag{7.32}$$

A similar analysis can be done for the state current, whereby

$$w_{32} = \frac{C_{\text{in}}}{C_T}(1 + \Delta_2) \tag{7.33}$$

The general form of the voltage V_A generated by the transistor M2 is

$$V_A = \frac{nU_T}{w_{22}} \ln\left(\frac{I_A}{I_s}\right) - \frac{w_{24}}{w_{22}} V_x \tag{7.34}$$

where w_{24} and w_{22} are the relative weights from every input to the FG in transistor M2 as shown in Fig. 7.5. w_{22} takes into account the contribution of the gate to drain coupling capacitance. Hence, the current provided to the integrating capacitance is then a function of V_x, and I_x:

$$I \propto e^{-(1+\Delta_1)/(1+C_{\text{GD}}^{(M2)}/C_{\text{in2}}+\Delta_3)(C_{\text{in}}/C_T)V_x} \propto (I_x)^{-((1+\Delta_4)/(1+C_{\text{GD}}^{M2}/C_{\text{in2}})(C_{\text{in}}/C_{\text{in4}}))} \tag{7.35}$$

where C_{in2} and C_{in4} are the unit capacitors in transistors M2 and M4, respectively, and $w_{24} = C_{\text{in2}}/C_T$, $w_{22} = (C_{\text{in2}}/C_T)(1 + C_{\text{GD}}^{(M2)}/C_{\text{in2}} + \Delta_3)$, $w_{43} = C_{\text{in4}}/C_T$ and $\Delta_4 = \Delta_2 - \Delta_1$. Therefore x_2 would be

$$x_2 = (1 + \Delta_4)\Big/\left[\frac{C_{\text{in4}}}{C_{\text{in}}}\left(1 + \frac{C_{\text{GD}}^{(M2)}}{C_{\text{in2}}}\right)\right] \tag{7.36}$$

Figure 7.14 and Fig. 7.15 show the THD versus different parameters[74], depending upon the amplitude of the input signal for $I_{\text{inDC}} = I_A = 50$ nA. The main sources of distortion are the gate to drain parasitic capacitance in transistor M2 and the mismatch between the input capacitance in the transistor providing current to the integrating

---

[74] Those that affected the most the value of the THD. It was studied for different ones, and also for different variation ranges in x_i, and these were found to be the ones that affected the THD the most.

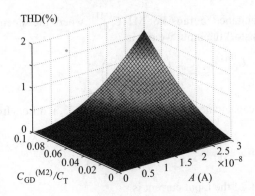

Figure 7.14   THD versus the gate to drain capacitance in M2, and the input amplitude

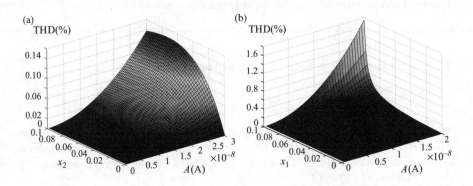

Figure 7.15   (a) THD versus the mismatch between the input capacitances affecting I_x eq. (7.36), and the input amplitude. (b) THD versus the mismatch between the input capacitances affecting I_{in} eq. (7.32), and the input amplitude

capacitance, and the input transistor, Fig. 7.14 and Fig. 7.15(b), respectively. The first source of distortion can be controlled by dimensioning the input capacitors in such a way that the relative weight C_{GD}/C_T is small enough compared with C_{in}/C_T. The second source of THD is much more technology dependent, but for typical mismatch values of 8 per cent and amplitudes below 20 nA it should be well below 1 per cent.

The previous analysis was performed under the assumption that both differential branches were completely symmetrical. Should this be true the even order harmonics would cancel out. However, in reality, mismatch is going to cause asymmetry and this in its turn might generate even-order harmonics comparable in magnitude to the odd-order ones. Mismatch can be included in the previous study by assuming that x_1 and x_2 are different for the positive and negative branches: $x_{1(n,p)}$, $x_{2(n,p)}$.

Also, the mismatch affecting both blocks gains can be taken into account as different coefficients normalising the amplitudes. Hence, all the b_i parameters will be split into two terms, b_{in} and b_{ip}, and the expressions for the two dominant distortion harmonics, second (H_{D2}) and third one (H_{D3}), are as follows:

$$H_{D2} = [(b_{1p}b_{3p}b_{6p} - b_{0p}b_{2p}b_{8p} + b_{0p}b_{4p}b_{6p})$$
$$- (b_{1n}b_{3n}b_{6n} - b_{0n}b_{2n}b_{8n} + b_{0n}b_{4n}b_{6n})]^2$$
$$+ \left[b_{2p}b_{1p}b_{7p} + \frac{b_{1p}b_{2p}b_{9p} + b_{1p}b_{4p}b_{7p} + b_{0p}b_{3p}b_{9p} + b_{0p}b_{5p}b_{7p}}{2} \right.$$
$$- b_{2n}b_{1n}b_{7n} - \frac{b_{1n}b_{2n}b_{9n} + b_{1n}b_{4n}b_{7n} + b_{0n}b_{3n}b_{9n} + b_{0n}b_{5n}b_{7n}}{2}$$
$$\left. + b_{0p}b_{3p}b_{7p} - b_{0n}b_{3n}b_{7n} \right]^2 \tag{7.37}$$

$$H_{D3} = ((b_{0p}b_{5p}b_{6p} - b_{0n}b_{5n}b_{6n} + b_{1p}b_{4p}b_{6p} - b_{1n}b_{4n}b_{6n} - b_{1p}b_{2p}b_{8p}$$
$$+ b_{1p}b_{2p}b_{8p})/2)^2 + ((b_{0p}b_{9p}b_{2p} - b_{0n}b_{9n}b_{2n} - b_{4p}b_{0p}b_{7p}$$
$$+ b_{4n}b_{0n}b_{7n} - b_{1p}b_{3p}b_{7p} + b_{1n}b_{3n}b_{7n})/2)^2 \tag{7.38}$$

And the main one is as follows:

$$H_{D1} = 16(b_{0p}b_{3p}b_{6p} - b_{0n}b_{3n}b_{6n} + b_{1p}b_{2p}b_{6p} - b_{1n}b_{2n}b_{6n}$$
$$+ \frac{3}{4}(b_0b_5b_6 + b_1b_4b_6 - b_1b_2b_8) + b_{1p}b_{2p}b_{8p} - b_{1n}b_{2n}b_{8n}$$
$$+ \frac{3}{4}(b_{0p}b_{5p}b_{6p} - b_{0n}b_{5n}b_{6n} + b_{1p}b_{4p}b_{6p} - b_{1n}b_{4n}b_{6n} - b_{1p}b_{2p}b_{8p}$$
$$+ b_{1n}b_{2n}b_{Sn}))^2 + 16(b_{0p}b_{2p}b_{7p} - b_{0n}b_{2n}b_{7n} + b_{4p}b_{0p}b_{7p} - b_{4n}b_{0n}b_{7n}$$
$$+ \frac{3}{4}(b_{0p}b_{9p}b_{8p} - b_{0n}b_{9n}b_{8n} - b_{4p}b_{0p}b_{7p} + b_{4n}b_{0n}b_{7n} - b_{1p}b_{3p}b_{7p}$$
$$+ b_{1n}b_{3n}b_{7n}) + b_{1p}b_{3p}b_{7p} - b_{1n}b_{3n}b_{7n})^2 \tag{7.39}$$

Hence, the second-order and third-order THD terms will be

$$HD2 = 100\sqrt{\frac{H_{D2}}{H_{D1}}} \tag{7.40}$$

$$HD3 = 100\sqrt{\frac{H_{D3}}{H_{D1}}} \tag{7.41}$$

Figure 7.16 and Fig. 7.17 represent these equations. Figure 7.16(a) shows the distortion caused by the second-order harmonic when the ratios between the capacitances that realise the exponents for the currents in the mathematical law are different

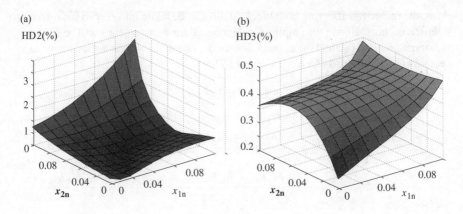

Figure 7.16   HD2 and HD3 versus x_{1n} and x_{2n}, when $x_{1p} = x_{2p} = 0.05$ and the input amplitude is 20e-9A

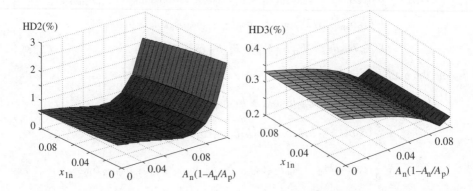

Figure 7.17   HD2 and HD3 versus x_{1n} and the mismatch affecting the gains (being $x_{1p} = x_{2p} = -0.05$, and $x_{2n} = 0$)

from one and also different from each other. The graph has been obtained for $x_{1p} = x_{2p} = 5$ per cent and an input amplitude of 20e-9A. The ratio between the third-order and first-order harmonic is represented in Fig. 7.16(b). It can be observed how under certain conditions the second-order harmonic may be larger than the third-order one. In Fig. 7.17 $x_{2p} = x_{1p} = 0.05$, $x_{2n} = 0$ and the y axis represents any mismatch affecting the positive and negative gains (A_p and A_n play the role of a normalised A taking into account different gains for the positive and negative differential sides in the integrator).

The main conclusion that can be extracted from the previous analysis is that the fundamental source of distortion is the mismatch between the same parameters for the two different single sides of the integrator. The second-order harmonic originated because of this could even be larger than the third-order one. The most

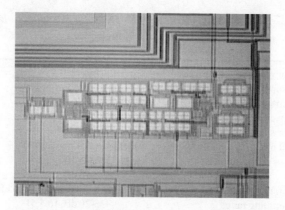

Figure 7.18   Fabricated prototype layout

critical parameters are those related to the integrator time constant (included under the form of different gains for the positive and negative parts, A_p and A_n). They could change mainly because of different I_s for the transistors, which is very usual in weak inversion. In any case, all these figures represent critical situations. In general, the typical values of mismatch for, for example, capacitances are going to be well below 10 per cent[75].

7.5   A design example

This section illustrates the previously developed theory with some experimental results obtained from a prototype design in a 0.35 μm technology ((AMS-CSX) [217]). The filter parameters obtained from eq. (7.14) and eq. (7.15) are as follows:

$$\omega_o = \sqrt{\omega_{o2}\omega_{o4}} \tag{7.42}$$

$$Q = \frac{\sqrt{\omega_{o2}\omega_{o4}}}{\omega_{o1}} \tag{7.43}$$

$$H_{LP}(0) = \frac{\omega_{o3}}{\omega_{o2}} \tag{7.44}$$

$$H_{BP}(\omega_o) = \frac{\omega_{o3}}{\omega_{o1}} \tag{7.45}$$

The filter operating at 1 V consumes less than 2 μW of power (this is without taking into account the expander block, as its power will vary depending on the current amplification required at the output). The filter performance and design parameters are summarised in Table 7.2. The layout is shown in Fig. 7.18. Figures 7.19–7.24

---

[75] And also the values of mismatch plus parasitic capacitances.

Table 7.2 *Summary of the 2nd-order filter performance and design parameters*

Technology parameters	$V_{Tn} = 0.45$ V $V_{Tp} = -0.62$ V
Voltage supply	1 V
Maximum power	2 μW
Input capacitances	58 fF
Transistor sizes (μm/μm)	5/0.8
HD3@70 mV$_{pp}$($I_{in} = 43$ nA$_{pp}$), $f_o = 500$ Hz	<0.4 per cent
DR	>45 dB
PSRR+	>40 dB
f_o programming range	>60 dB, [40 Hz,40 kHz]
Q programming range	>20 dB, [0.5,5]
Gain programming range	>40 dB
Area	175 μm × 265 μm

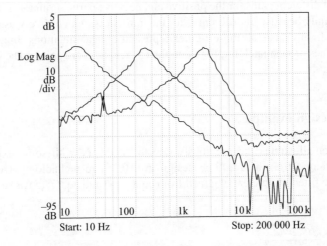

Figure 7.19 *Programming of the cut-off frequency in the bandpass output*

show experimental examples of programming. The transfer function of the lowpass realisation exhibits a peak which is caused by additional roots due to the parasitic capacitances. Nevertheless, the peak is at least 30 dB below the gain in the passband of the filter and so is not critical. It originates from the parasitic capacitances and is not easy to deduce analytically, although it is normal in weak inversion realisations. The measured Signal to Noise Ratio is shown in Fig. 7.25. Results of THD and the H$_{D3}$ (which is directly related to the IM3 as seen in previous chapters) are shown in Fig. 7.26 and Fig. 7.27. Figure 7.27 stresses the importance of the second-order

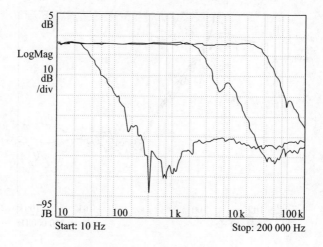

Figure 7.20   Programming of the cut-off frequency in the lowpass output

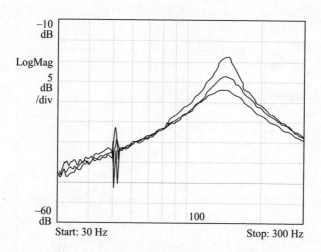

Figure 7.21   Programming of the quality factor in the bandpass output

harmonic previously discussed in the theoretical study. In the figure the latter is only 32 dB beneath the fundamental tone.

7.6   Summary and conclusions

This chapter described how to use FGMOS transistors to design internally nonlinear and externally linear logarithmic filters. It has been shown how very compact

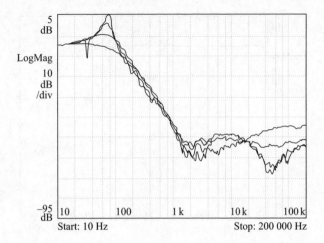

Figure 7.22   Programming of the quality factor in the lowpass output

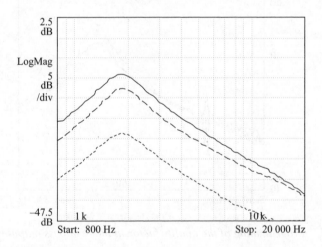

Figure 7.23   Programming of the gain in the bandpass output

realisations can be designed by using the inputs in the transistor for different purposes, such as tuning, signal processing, biasing and control of the common mode. Because of this there is no need of stacking transistors to implement translinear loops, which simplifies the design under the low-voltage constraint apart from having other added advantages that have also been explained along the chapter. The main

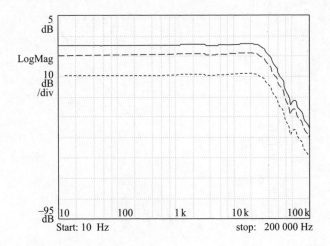

Figure 7.24   Gain programming in the lowpass filter

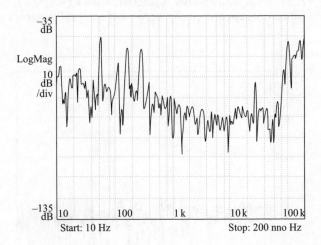

Figure 7.25   1/Signal to noise ratio

disadvantage of these realisations is related to the parasitic couplings to the FGs in the FGMOS devices that together with the mismatch increase the distortion levels. In general though this kind of topology performs very well in low frequency, low accuracy applications in which the main design constraints are power and low supply voltage.

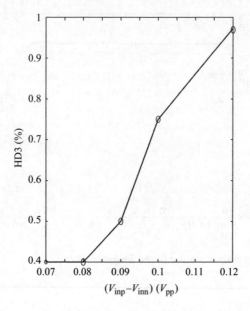

Figure 7.26   *HD3 versus the amplitude of a 500 Hz input signal for a bandpass filter with unity gain and quality factor, and a cut-off frequency of 500 Hz*

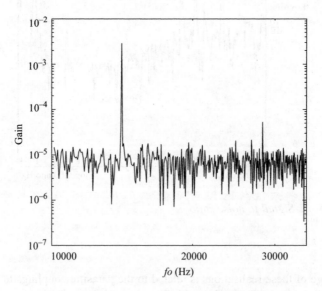

Figure 7.27   *Experimental second order harmonic for a low pass filter with a cut-off frequency of 40 kHz, quality factor of 1, and an input signal of 200 mV_pp*

Notation

α	Coefficient of the state variable in the state-space equation of a linear lossy integrator (eq. (7.1))
α'	Modified expression of α if mismatch and C_{GD} are taken into account (eq. (7.26))
Δ_i for $i = [1,4]$	Parameter accounting for the percentage of mismatch between different capacitances
η	Input coefficient in the state-space equation of a linear lossy integrator (eq. (7.1))
η'	Modified expression of η if mismatch and C_{GD} are taken into account (eq. (7.26))
ω_o	Biquad cut-off frequency (rad/s. Eq. (7.42))
ω_{oi} for $i = [1,4]$	Constants in the state-space equations of the second-order filter prototype (eqs. (7.14) and (7.15))
A	I_{in} amplitude
A_n	Normalised A for the n side of the fully differential integrator taking into account different gains
A_p	Normalised A for the p side of the fully differential integrator taking into account different gains
A_1	I_x amplitude
b_i for $i = [1,9]$	See Table 7.2
b_{in}	b_i parameters for the n side of the fully differential integrator when mismatch is considered
b_{ip}	b_i parameters for the p side of the fully differential integrator when mismatch is considered
C_{GD}^{Mi}	Gate to drain capacitance for transistor Mi (Fig. 7.5)
C_{in}/C_T	Weight of the effective input w
C_{in2}	Unit capacitance in transistor M2 (For study of mismatch. Fig. 7.5)
C_{in4}	Unit capacitance in transistor M4 (For study of mismatch. Fig. 7.5)
f_o	Biquad cut-off frequency (Hz)
Q	Biquad quality factor (eq. (7.43))
G_{ex}	Gain of the expander block as given by eq. (7.20)
$H_{LP}(0)$	Lowpass filter DC gain (eq. (7.44))
$H_{BP}(\omega_o)$	Lowpass filter DC gain (eq. (7.44))
I_{1n}	Single state variable related to the bandpass output (eq. (7.14))
I_{1p}	Single state variable related to the bandpass output (eq. (7.14))
I_{3n}	Single state variable related to the lowpass output (eq. (7.14))
I_{3p}	Single state variable related to the lowpass output (eq. (7.14))

I_A	Bias current in the divider block (Fig. 7.3), in the input stage (Fig. 7.9) and in the CMFB (Fig. 7.11)
I_B	Current related to the parameter α in the integrator (eq. (7.11), Fig. 7.4)
I_{in}	Input current (eq. (7.6))
$I_{in_{DC}}$	I_{in} DC component
I_{inn}	Single input current in one of the sides of the fully differential integrator (eq. (7.14))
I_{inp}	Single input current in one of the sides of the fully differential integrator (eq. (7.14))
I_{jn} for $j = 1, 3$	Output voltage in the expander block
I_{jp} for $j = 1, 3$	Output voltage in the expander block
I_{out}	Output current in the integrator in Fig. 7.4
I_x	State current (eq. (7.5))
$I_{x_{DC}}$	I_x DC component
I_{xn}	See I_{outn} and I_x
I_{xp}	See I_{outp} and I_x
I_{out1}	Output of the NLB (Fig. 7.3)
I_{outj}	$I_{jp} - I_{jn}$ for $j = [1, 3]$. Output of the expander block (Fig. 7.8, eq. (7.23))
I_{outn}	I_{xn}. Single output of the expander block in the fully differential integrator (Fig. 7.4)
I_{outp}	I_{xp}. Single output of the expander block in the fully differential integrator (Fig. 7.4)
I_{out1i}	Output of a generic NLB with an extra input connected to V_{tuni}
K	$\eta k_{in}/k_y$ (eq. (7.4))
k_{in}	Multiplicative coefficient in the equation relating x_{in} and u (eq. (7.2))
k_y	Multiplicative coefficient in the equation relating x and y (eq. (7.2))
(n, p)	Subscript used to simultaneously refer to devices in the negative and positive side of the fully differential integrator
u	Variable related to x_{in} in a logarithmic way (eq. (7.2))
V_{1n}	Voltage across one of the terminals of the integrating capacitance associated with the bandpass output (eq. (7.16))
V_{1p}	Voltage across one of the terminals of the integrating capacitance associated with the bandpass output (eq. (7.16))
V_{3n}	Voltage across one of the terminals of the integrating capacitance associated with the lowpass output (eq. (7.16))
V_{3p}	Voltage across one of the terminals of the integrating capacitance associated with the lowpass output (eq. (7.16))
V_A	Voltage generated by I_A in the NLB (Fig. 7.3)
V_{An}	V_A voltage in the n side of the fully differential integrator
V_{Ap}	V_A voltage in the p side of the fully differential integrator
V_B	Constant voltage input in the expander block (eq. (7.5))

V_{B1}	Voltage input in one of the two expanders in Fig. 7.7
V_{B2}	Voltage input in one of the two expanders in Fig. 7.7
V_{Bin}	Constant input in the compressor block (eq. (7.7))
V_{cm}	Constant voltage applied to certain FGMOS transistors as explained in Section 7.2
V_e	Constant voltage in the expander block (Fig. 7.8)
V_{in}	Voltage generated by I_{in} in the compressor block (eq. (7.7))
V_{inn}	Voltage generated by I_{inn} in the compressor block (eq. (7.7))
V_{inp}	Voltage generated by I_{inp} in the compressor block (eq. (7.7))
V_{jn} for $j = [1, 3]$	Input voltage in the fully differential expander block
V_{jp} for $j = [1, 3]$	Input voltage in the fully differential expander block
V_{outcm}^n	Negative output of the CMFB associated with the bandpass output
$V_{outcm}'^n$	Negative output of the CMFB associated with the lowpass output
V_{outcm}^p	Positive output of the CMFB associated with the bandpass output
$V_{outcm}'^p$	Positive output of the CMFB associated with the lowpass output
V_{tuni}	Tuning voltage that controls the value of ω_{oi} (for $i = [1, 4]$; eq. (7.21))
V_x	Voltage that generates I_x in the expander block (eq. (7.5))
V_{xn}	Voltage that generates I_{xn} in the expander block (Fig. 7.4)
V_{xp}	Voltage that generates I_{xp} in the expander block (Fig. 7.4)
w	Weights of the FGMOS transistors in the divider block
w_B	Weight of input V_B in the expander block (eq. (7.5)). Equal to w in subsequent sections
w_{Bin}	Weight of input V_{Bin} in the compressor block (eq. (7.6)). Equal to w in subsequent sections
w_D	C_{GD}/C_T
w_{ij}	Input weight associated with capacitance in transistor Mi connected to transistor Mj (Fig. 7.5)
w_{in}	Weight of input V_{in} in the compressor block (eq. (7.6)). Equal to w in subsequent sections
w_x	Weight of input V_x in the expander block (eq. (7.5)). Equal to w in subsequent sections
x	State variable in the equation of a linear lossy integrator (eq. (7.1))
x_1	Exponent of the input current in the state-space equation when mismatch is considered within the same branch only (eq. (7.26))
$x_{1(n,p)}$	Parameter used to refer simultaneously to x_1 for the n and p side of the integrator when mismatch makes them different

x_2	Exponent of the input current in the state-space equation when mismatch is considered within the same branch only (eq. (7.26))
$x_{2(n,p)}$	Parameter used to refer simultaneously to x_2 for the n and p side of the integrator when mismatch makes them different
x_{in}	Input variable in the equation of a linear lossy integrator (eq. (7.1))
x_{inn}	Single input variable for a fully differential implementation in eq. (7.12)
x_{inp}	Single input variable for a fully differential implementation in eq. (7.12)
x_n	Single state variable for a fully differential implementation in eq. (7.12)
x_p	Single state variable for a fully differential implementation in eq. (7.12)
y	Variable related to x in a logarithmic way (eq. (7.2))

Chapter 8

Low power digital design based on the FGMOS threshold gate

8.1 Introduction

This chapter introduces a completely different application for the FGMOS transistors, in digital circuit design. The device can be used to implement digital functions in a much more compact way, which results in the desired reduction of power. The methodology followed for this consists of the following steps: 1. Design the digital functions at the system level using threshold gates. 2. Design the corresponding threshold gates at the circuit level using FGMOS transistors. 3. Design the whole system at the circuit level using the threshold gates developed in 2.

Threshold Gates are digital circuits whose functionality is based on the so called majority or threshold decision principle. The threshold decision principle states that the value of the signal at the circuit output depends on whether the arithmetic sum of values at the inputs exceeds or falls below a threshold value. This general definition includes as special cases conventional logic gates, such as AND and OR.

For many years, logic circuit design based on threshold gates has been considered as an alternative to the traditional logic gate design procedure. The threshold gate intrinsically implements a complex function. If, instead of using standard logic gates, the threshold gate is used as main block of the system, the number of gates as well as the gate levels can be reduced. However, the growing interest in threshold logic is mainly due to the recent theoretical results that show how bounded level networks of threshold gates can implement functions that would require unbounded level networks of standard logic gates. Important functions such as multiple-addition, multiplication, division or sorting can be implemented by polynomial-size threshold circuits with a small constant depth.

Threshold gates are very attractive for low power design as their use can simplify the topology of digital systems which normally results in a power reduction. This chapter shows how a threshold gate can be built in a very compact form using the

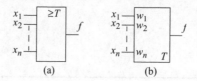

Figure 8.1   (a) Standard and (b) Non-standard symbol for the threshold gate

FGMOS (Neuron-νMOS[76]) transistor. This threshold gate is used to design several important digital blocks that are commonly used in more complex digital systems. The figure of merit ***power*** $\times$ ***delay*** for all presented blocks is shown and compared with the figure of merit of conventional architectures, confirming the potential of the νMOS device for low power digital design.

8.2   Threshold gates

An *n*-input threshold gate (TG) is defined as a logic gate with *n* inputs, x_i, $(i = 1, \ldots, n)$, which can take values 0 and 1, and for which there is a set of $(n + 1)$ real numbers $w_1, w_2, \ldots, w_n$ and T, called weights and threshold, respectively, such that the output of the gate is

$$f = \begin{cases} 1 & \text{for } \sum_{i=1}^{n} w_i x_i \geq T \\ \\ 0 & \text{for } \sum_{i=1}^{n} w_i x_i < T \end{cases} \quad (8.1)$$

A function represented by the output of a TG, denoted by $f(x_1, x_2, \ldots, x_n)$, is called a threshold function. The set of weights and threshold can be written in a more compact vector notation as $[w_1, w_2, \ldots, w_n; T]$.

Figure 8.1(a) shows the IEEE standard symbol for a TG with all the input weights equal to 1 and threshold T. There is no standard symbol for a threshold gate with weights different from 1, but it can be built by tying together several inputs (a weight of w_i for the input x_i can be obtained by connecting x_i to w_i gate inputs). Instead of that the non-standard symbol for TGs with generic weights is used (Fig. 8.1(b)).

A threshold gate can be defined with positive and negative weights as well as with positive and negative thresholds. However, in order to facilitate the physical implementation it is useful to have an alternative TG description only with the positive parameters. This can easily be obtained by taking into account the elementary relations of threshold functions shown in Fig. 8.2. The arrow labelled with (1) means that if $f(x_1, x_2, \ldots, x_n)$ is a threshold function defined by

---

[76] Both names will be used indistinctly from now on. The second one is commonly used in digital design, whereas the first is a much more general nomenclature.

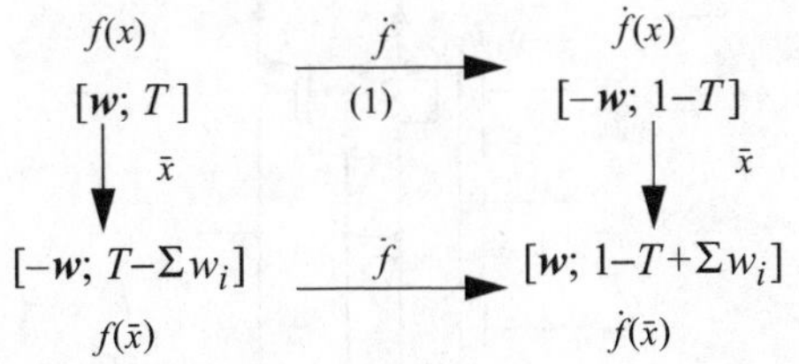

$$f(x) \qquad \dot{f} \qquad \dot{f}(x)$$
$$[w;\,T] \qquad (1) \qquad [-w;\,1-T]$$

$$\bar{x} \qquad\qquad\qquad \bar{x}$$

$$[-w;\,T-\Sigma w_i] \qquad \dot{f} \qquad [w;\,1-T+\Sigma w_i]$$
$$f(\bar{x}) \qquad\qquad\qquad \dot{f}(\bar{x})$$

Figure 8.2   Elementary relations of threshold functions

$[w_1, w_2, \ldots, w_n; T]$, then its complement $\dot{f}(x_1, x_2, \ldots, x_n)$ is also a threshold function defined by $[-w_1, -w_2, \ldots, -w_n; 1-T]$. If a function can be realised as a threshold element, then by selectively complementing the inputs, it is possible to obtain a realisation using a threshold element with only positive weights.

Once the architecture has been chosen, the effectiveness of the system implementation is determined by the availability, cost and capabilities of its basic building block. Various interesting circuit concepts have been reported for developing standard-CMOS compatible threshold gates. They can be classified as static and dynamic realisations, depending upon the context in which they are used. The Neuron MOS (νMOS) threshold gate constitutes one of the most notable contributions to the implementation of static-CMOS threshold gates [227–235].

8.3   The νMOS threshold gate

The realisation of very large logic circuits using νMOS transistors has been found to be a very promising alternative to conventional approaches [50–52,58]. This is mainly due to the enhancement of the functional capability of a basic transistor element which effectively reduces the complexity of the overall circuit.

Logic design techniques for implementation of νMOS circuits were introduced by Shibata [50–52], but their usefulness was limited because, in general, they led to complex circuit configurations which required handling 2^n logic states for an n-input logic function. This imposed stringent constraints on process tolerances which are not realisable by present technologies, even in cases when n is small. A different and more powerful approach for designing using νMOS takes advantage of the fact that the functionality of νMOS circuits is closely related to the functionality of a TG, Fig. 8.3 [235–237]. The rest of this section explains the implementation of the TG with νMOS transistors, first at a theoretical level and then briefly discussing several practical considerations.

8.3.1   Theory

The simplest νMOS-based threshold gate (νMOS-TG) is a complementary inverter using both p- and n-channel FGMOS devices, followed by a normal CMOS inverter. The schematic of this TG is shown in Fig. 8.3. The FG is usually shared by both the

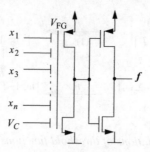

Figure 8.3   vMOS Threshold Gate (vMOS TG)

pMOS and the nMOS transistors, although this does not have to be always the case. There are several input gates $(x_1, x_2, \ldots, x_n)$, corresponding to the TG inputs plus an extra input (indicated by V_c in the figure) for logic threshold adjustment (it will be explained later). Neglecting the parasitic capacitances, and ignoring for now the extra control input, the voltage at the FG is

$$V_{FG} = \left(\sum_{i=1}^{n} C_i V_{xi} \right) \bigg/ C_T \tag{8.2}$$

where C_i is the input capacitance corresponding to the input x_i, C_T is the total capacitance seen by the FG and V_{xi} is the voltage applied to the input gate x_i. When V_{FG} becomes higher than the inverter switching threshold voltage, U_{inv}, the output switches to logic 1.

A vMOS TG has digital entries, i.e. $V_{x_i} = x_i V_{DD}$, where V_{DD} is the power supply voltage and $x_i \in \{0, 1\}$. There is a relation between the above equation for V_{FG} and the definition of the TG given by eq. (8.1). Namely, the weighted summation in the TG, $\sum_{i=1}^{n} w_i x_i$, can be implemented by the capacitive network in the vMOS device:

$$\left(\sum_{i=1}^{n} C_i x_i V_{DD} \right) \bigg/ C_T \tag{8.3}$$

The weight for each input is then proportional to the ratio between the corresponding input capacitance C_i and C_T:

$$w_i = C_i V_{DD} / C_T \tag{8.4}$$

Thus, design involves mapping the logical inequalities

$$\sum_{i=1}^{n} w_i x_i \geq T \tag{8.5}$$

$$\sum_{i=1}^{n} w_i x_i < T \tag{8.6}$$

to the electrical relationships given by eq. (8.7) and eq. (8.8) by means of capacitances sizing and tuning of the switching threshold of the inverter:

$$\left(\sum_{i=1}^{n} C_i x_i V_{DD} \right) \Big/ C_T > U_{inv} \tag{8.7}$$

$$\left(\sum_{i=1}^{n} C_i x_i V_{DD} \right) \Big/ C_T < U_{inv} \tag{8.8}$$

When the logical threshold T is not centred[77], the threshold voltage of the inverter also needs to be non-centred (near 0 or V_{DD}). This can be achieved using extra inputs. For example, let us assume a single control input c with capacitance $C_{Control}$. If V_c is applied to this control input c, the new V_{FG} is[78]

$$V_{FG} = \left(\sum_{i=1}^{n} C_i x_i V_{DD} \right) \Big/ C_{tot} + (C_{Control} V_c)/C_T \tag{8.9}$$

$$C_T = \sum_{i=1}^{n} C_i + C_{Control} \tag{8.10}$$

From the point of view of the TG, a comparison is performed between (8.3) and

$$U_{inv} - (C_{Control} V_c)/C_T \tag{8.11}$$

Thus, the effective threshold voltage of the inverter can be modified. In practical digital design analog voltages are avoided and so the role of the analog extra input V_c is realised by a set of digital inputs with the appropriate coupling capacitances.

In a real design of a υMOS TG second-order effects also need to be considered, although they were not included in the previous equations for the sake of clarity. The main issue that should be taken into account in the circuit in Fig. 8.3 is related to the coupling effects when switching on the power supply. Capacitances C_{GD}, C_{GS} and C_{GB} are responsible for the effects, causing additional terms in V_{FG}. However, these can be minimised either by sizing adequately the input capacitors and transistors or choosing the switching threshold, accordingly.

8.3.2 Practical design aspects

The previous section described the theory behind the implementation of TGs with υMOS devices. However, practical design considerations also have to be taken into

---

[77] i.e. T is far from $(\sum_{i=1}^{n} w_i)/2$.
[78] Neglecting the parasitic couplings.

account and the following issues must be solved:

1. *Signal regeneration*. On one hand, these TGs exhibit reduced noise margins as a consequence of the circuit electrical operation. On the other hand, some input combinations can produce output voltages different from V_{DD} or ground. These two considerations are specially critical when logic networks are built up of interconnecting νMOS-TGs.

2. *Electrical simulation*. The electrical simulation problem has been already addressed in Chapter 2; so, it will not be considered here again. The simulation problems can be avoided by using the ITA method described in detail in Section 2.4.

8.4 νMOS Threshold gates applications

The implementation of digital circuits using νMOS TGs is illustrated in this section with several examples. These focus on the design of low power-high speed blocks. These examples can be classified into combinational and sequential circuits. The first part of the section is divided into arithmetic combinational blocks, specifically TG based adders using floating-gate circuits [238] and νMOS-based compressor designs [239–240]. It also describes a sorting network [241] that is used as the main functional cell in the implementation of arithmetic circuits afterwards [242–244]. The second part describes, characterises and illustrates with experimental results the design of a sequential block: the multi-input Muller C-element, as an example of the behaviour of the νMOS TG in a configuration with feedback [124].

8.4.1   *Threshold logic based adders using floating-gate circuits*

Binary addition is an arithmetic operation widely used in many fields. It is also essential for any system containing circuitry for subtraction, multiplication and division [245–248]. Often it is the limiting element in the speed operation of such systems, and therefore implementing high-performance adders is of utmost importance. Optimisation of adders can be realised either at the logic or at the circuit level. Logic optimisation generally comes from rewriting the logic equations which define the adder in such a way that their implementation results in a cheaper or faster realisation. Circuit optimisation is based on considering some of the realisation electrical aspects such as topology and transistor sizing.

A bottleneck in the design of adders is the carry generation. N-bit ripple-carry adders have a propagation delay linear with N and these are only useful for addition of numbers with a relatively small word length. Nowadays, adders for a word length up to 128 are required which makes the use of N-bit ripple-carry adders impractical. One of the most common techniques to avoid this problem, and to design very fast logic adders, uses the carry lookahead principle. The latter allows the optimisation of the adder at a logic level resulting in implementations which effectively eliminate the undesired ripple effect and exhibit a constant time delay. However, the physical implementation of these carry lookahead adders can become impractical because of the very high fan-ins required. To solve this problem a hierarchical approach is usually taken, and a logarithmic time delay is so obtained.

This section will show how to implement a very fast carry lookahead by using a design style based on TGs implemented as νMOS circuits. The technique is very effective to cut-off the time delay and to increase the maximum operating frequency of these adders.

8.4.1.1   Carry lookahead adders

Let $\mathbf{A} = A_{n-1}, \ldots, A_1, A_0$ and $\mathbf{B} = B_{n-1}, \ldots, B_1, B_0$ be the augend and addend inputs to an n-bit adder, C_{i-1} the carry input to the i-th bit position and C_{-1} the corresponding input to the least significant position. In a carry lookahead adder, the carry-out of each stage, C_i, is directly defined as a function of $A_i, A_{i-1}, \ldots, A_0, B_i, B_{i-1}, \ldots, B_0$ and C_{-1}. By defining two auxiliary functions, the carry generate G_i and the carry propagate P_i functions, the sum and carry outputs of the i-th stage can be obtained as

$$S_i = P_i \oplus C_{i-1} \qquad G_i = A_i \cdot B_i$$
$$C_i = G_i + P_i \cdot C_{i-1} \qquad P_i = A_i \oplus B_i \tag{8.12}$$

These equations mean that if all the carry inputs $C_{n-2}, \ldots, C_1, C_0, C_{-1}$ are available simultaneously, then all the sum bits S_i for $i = n - 1, \ldots, 1, 0$ can be generated in parallel. By expanding the recursive formula for C_i, the set of eqs. (8.13) for the carries can be obtained, which can be realised in two logic levels:

$$C_0 = G_0 + P_0 \cdot C_{-1}$$
$$C_1 = G_1 + P_1 \cdot C_0 = G_1 + P_1 \cdot (G_0 + P_0 \cdot C_{-1})$$
$$= G_1 + P_1 \cdot G_0 + P_1 \cdot P_0 \cdot C_{-1}$$
$$C_2 = G_2 + P_2 \cdot C_1 = G_2 + P_2 \cdot (G_1 + P_1 \cdot C_0)$$
$$= G_2 + P_2 G_1 + P_2 P_1 G_0 + P_2 P_1 P_0 C_{-1}$$
$$\dots\dots\dots\dots\dots\dots\dots\dots\dots \tag{8.13}$$
$$C_k = G_k + P_k \cdot G_{k-1} = G_k + P_k \cdot (G_{k-1} + P_{k-1} \cdot C_{k-2})$$
$$= G_k + P_k G_{k-1} + P_k P_{k-1} G_{k-2} + \cdots$$
$$+ P_k P_{k-1} \cdots P_2 P_1 G_0 + P_k \cdots P_1 P_0 C_{-1}$$
$$\dots\dots\dots\dots\dots\dots\dots\dots\dots\dots$$
$$C_{n-1} = G_{n-1} + P_{n-1} G_{n-2} + \cdots + P_{n-1} \cdots P_1 P_0 C_{-1}$$

The final sum bits are available after a delay Δ equal to

$$\Delta = \Delta_{\text{PG}} + \Delta_{\text{CLA}} + \Delta_{\text{S}} \tag{8.14}$$

where Δ_{PG} is the circuit delay due to the carry generate/propagate unit generating the P_i and G_i signals, Δ_{CLA} is the delay of the CLA producing the C_i signals and Δ_{S} corresponds to the generation of the S_i sum signals (summation unit). Hence, for example, in a 4-bit realisation, under the simplifying assumption that a gate delay is

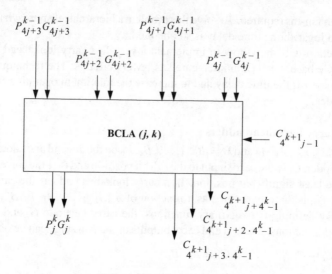

Figure 8.4   4-bit block carry lookahead (BCLA) unit

constant and equal to δ, the final sum bit is available after 8δ:3δ (corresponding to the circuit delay to generate the P_i and G_i signals, Δ_{PG} plus 2δ (for Δ_{CLA}) plus 3δ (for Δ_S). However, direct calculation of carries beyond 4 bits becomes impractical because of the very high fan-ins required in the implementation, and a hierarchical approach is needed [249].

The hierarchical solution uses the so-called block carry lookahead (BCLA) unit. Each one of such units computes its own 'group' carry propagate and generate from the signals coming on it. Figure 8.4 shows the 4-bit BCLA unit corresponding to a generic j-th column and k-th row in an adder tree implementation.

Equations defining the outputs of this BCLA are the carry signals $C_{4^{k+1}j+4^k-1}, C_{4^{k+1}j+2\cdot4^k-1}, C_{4^{k+1}j+3\cdot4^k-1}$, the group propagate signal P_j^k, which determines whether a carry into the block would result in a carry out of the block, and the group generate signal G_j^k, corresponding to the condition that the carry generated out of the most significant position of the block was originated within the block itself:

$$C_{4^{k+1}j+4^k-1} = G_{4j}^{k-1} + P_{4j}^{k-1} \cdot C_{4^{k+1}j-1}$$

$$C_{4^{k+1}j+2\cdot4^k-1} = G_{4j+1}^{k-1} + G_{4j}^{k-1} \cdot P_{4j+1}^{k-1} \cdot P_{4j+1}^{k-1} \cdot P_{4j}^{k-1} \cdot C_{4^{k+1}j-1}$$

$$C_{4^{k+1}j+3\cdot4^k-1} = G_{4j+2}^{k-1} + G_{4j+1}^{k-1} \cdot P_{4j+2}^{k-1} + G_{4j}^{k-1} \cdot P_{4j+2}^{k-1} \cdot P_{4j+1}^{k-1}$$
$$+ P_{4j+2}^{k-1} \cdot P_{4j+1}^{k-1} \cdot P_{4j}^{k-1} \cdot C_{4^{k+1}j-1} \tag{8.15}$$

$$G_j^k = G_{4j+3}^{k-1} + G_{4j+2}^{k-1} \cdot P_{4j+3}^{k-1} + G_{4j+1}^{k-1} \cdot P_{4j+3}^{k-1} \cdot P_{4j+2}^{k-1}$$
$$+ G_{4j}^{k-1} \cdot P_{4j+3}^{k-1} \cdot + P_{4j+2}^{k-1} \cdot P_{4j+1}^{k-1}$$

$$P_j^k = P_{4j+3}^{k-1} \cdot P_{4j+2}^{k-1} \cdot P_{4j+1}^{k-1} \cdot P_{4j}^{k-1}$$

Let us assume an n-bit adder and a b-bit BCLA. The delay corresponding to this hierarchical solution is

$$\Delta = \Delta_{PG} + (\lceil \log_b n \rceil - 1) \cdot \Delta_{BCLA} + \Delta_{CLA} + \Delta_S \qquad (8.16)$$

where the additional term Δ_{BCLA} is the circuit delay of the block carry lookahead unit, $(\lceil \log_b n \rceil - 1)$ is the depth of the adder tree and the term Δ_{CLA} is now the delay of the final CLA unit.

The next section shows an alternative representation of the carry lookahead adder using TGs.

8.4.1.2   Threshold gate based implementation

Optimised high-performance adders can be obtained by resorting to threshold logic. The application of the threshold gate design methodology allows the reduction of the delay for the carries in a carry lookahead implementation. This is because it is possible to obtain the carry signals C_0, C_1, C_2 and 'group' carry propagate and generate functions as TGs whose weights and thresholds are as follows:

$$C_{4^{k+1}j+4^k-1}\left(C_{4^{k+1}j-1}, P_{4j}^{k-1}, G_{4j}^{k-1}\right) = [1,1,2;2]$$

$$C_{4^{k+1}j+2\cdot4^k-1}\left(C_{4^{k+1}j-1}, P_{4j}^{k-1}, G_{4j}^{k-1}, P_{4j+1}^{k-1}, G_{4j+1}^{k-1}\right) = [1,1,2,3,5;5]$$

$$C_{4^{k+1}j+3\cdot4^k-1}\left(C_{4^{k+1}j-1}, P_{4j}^{k-1}, G_{4j}^{k-1}, P_{4j+1}^{k-1}, G_{4j+1}^{k-1}, P_{4j+2}^{k-1}, G_{4j+2}^{k-1}\right)$$

$$= [1,1,2,3,5,8,13;13] \qquad (8.17)$$

$$G_j^k\left(G_{4j}^{k-1}, P_{4j+1}^{k-1}, G_{4j+1}^{k-1}, P_{4j+2}^{k-1}, G_{4j+2}^{k-1}, P_{4j+3}^{k-1}, G_{4j+3}^{k-1}\right)$$

$$= [1,1,2,3,5,8,13;13]$$

$$P_j^k\left(P_{4j}^{k-1}, P_{4j+1}^{k-1}, P_{4j+2}^{k-1}, P_{4j+3}^{k-1}\right) = [1,1,1,1;4]$$

These results are not specific for these carry signals. It can be proven that all the elements of the set of carry signals $C_0, C_1, \ldots, C_{n-1}$ are threshold functions which can be implemented in one logic level. Weights and threshold of the TGs obtained are related with the Fibonacci's number series in the following way:

$$C_j(C_{-1}, P_0, G_0, P_1, G_1, \ldots, P_{j-1}, G_{j-1}, P_j, G_j)$$

$$= [1,1,2,3,5,\ldots, F_{2j}, F_{2j+1}, F_{2j+2}, F_{2j+3}, F_{2j+3}] \qquad (8.18)$$

where F_k is the k-th Fibonacci's number, i.e. each F_k is the sum of the two adjacent Fibonacci's numbers on its immediate left, $F_k = F_{k-1} + F_{k-2}$. The total weight for the TG implementing C_j, $W_T(C_j)$, is given by

$$W_T(C_j) = 1 + 1 + 2 + \cdots + F_{2j+2} + F_{2j+3} = F_{2j+5} - 1 \qquad (8.19)$$

The availability of all the carry signals in only one level in a TG based implementation is theoretically important but of a limited practical interest since its usefulness depends on the physical availability of TGs implementing the total weight. This will be discussed in the next section. In general, there is a bound to the maximum total weight physically implementable on a TG, and so a hierarchical approach will be also needed. The practical difference with a traditional solution is that the BCLA unit can be implemented by one level of TGs, while it requires a two-level network when implemented with traditional gates. The delay for the *n*-bit adder, Δ^{TG}, when a hierarchical approach is followed becomes

$$\Delta^{TG} = \Delta_{PG} + (\lceil \log_b n \rceil - 1) \cdot \Delta^{TG}_{BCLA} + \Delta_{CLA} + \Delta_S \tag{8.20}$$

where Δ_{CLA} is the delay of the final CLA unit (the same two-level circuit for both approaches) and Δ^{TG}_{BCLA} is the circuit delay of the BCLA unit based on TGs. From equations (8.16) and (8.20) and assuming a delay of δ units per level, the difference between both delays is given by $(\lceil \log_b n \rceil - 1) \cdot \delta$; i.e.

$$\Delta = \Delta^{TG} + (\lceil \log_b n \rceil - 1) \cdot \delta \tag{8.21}$$

That is, the delay improves as *n* increases in an *n*-bit adder. To actually evaluate the reduction of the delay, it is necessary to exactly know the design style used in the implementation of the TGs. The next section analyses how the choice of technology affects the implementation of the carry lookahead adder.

8.4.1.3 Physical implementation

The feasibility of the TG approach discussed in Section 8.3 depends on the complexity of the TG that can be built with νMOS transistors. Results obtained from an extensive set of simulations showed that carry signals implemented by TGs up to a maximum number of bits equal to 4 ($b = 4$) can be built. If both 4-bit BCLAs, the traditional BCLAs and the νMOS based one, are designed, the delay of the νMOS implementation is around half of the delay of the traditional implementation. This reduction is of the greatest importance because this delay multiplies the depth of the network implementing the adder. The higher the number of bits of the adder the larger the reduction of the delay. This is illustrated in the following with the design of a 64-bit adder.

8.4.1.4 The 64-bit adder

An example of the application is the 64-bit adder shown in Fig. 8.5. The adder uses the 4-bit BCLA unit in Fig. 8.4. The carry generate/propagate unit and the summation unit are not shown, but they are a direct implementation of eq. (8.12). The 64-bit adder has two levels of 4-bit BCLAs and a final level with a 4-bit CLA; $j = 0, k = 0$ is the block at the right top position. The results of performance shown are obtained in a prototype designed and laid out in a 5 V 0.8 μm double poly CMOS process [214].

Correct operation under process, mismatch and ambient parameter variations are validated through extensive Monte Carlo HSPICE simulations of the extracted circuit. Transient characteristics and average power are measured on post-layout simulation results using typical device parameters at a supply voltage of 3 V. The worst case delay time is 5.6 ns and the power consumption is 19.6 mW at 100 MHz.

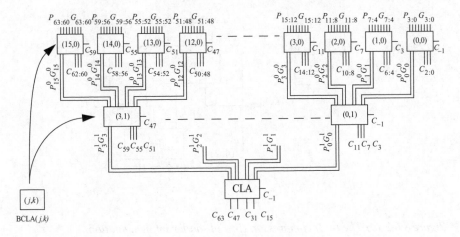

Figure 8.5   Core of a 64-bit adder using the carry lookahead principle

However, the intrinsic nature of the νMOS approach makes this power consumption very independent of the frequency.

In order to validate the proposed circuit, a comparison with another implementation is required. The 64-bit adder is also designed and laid out following a conventional approach (NAND gates were used in the two-level implementation of the BCLA) and with the same technological process. The worst case delay for this conventional design is over 11 ns and the power consumption at 50 MHz is equal to the power consumption of the TG-based implementation. Also, unlike the proposed design this one is very dependent on frequency.

8.4.2 νMOS-based compressor designs

Partial-product reduction circuits (compressors) are of capital importance in the design of high performance parallel multipliers. This section focuses on compressor design based on TGs implemented as νMOS circuits.

The dominant delay in the critical signal path of a parallel multiplier is due to the partial-product summation tree (PPST) [250]. A reduction of this delay is important in order to obtain high performance parallel multipliers. It is well known that it is hard to do layout for tree structures and hence large wiring channels are required. Full adders have been widely used as the basic building block but its 3:2 nature (it takes three inputs of the same weight and produces two outputs) makes it impossible to design a completely regular layout. When compressors are used as building blocks, more regular tree structures with fewer levels are obtained. Thus, PPSTs built with compressors instead of with full-adders have been preferred and so the design of high-performance compressors appears as a key problem in the design of high-performance multipliers.

An $(N, 2)$ compressor is defined to be a combinational circuit which operates in a single column of the PPST taking as input N bit signals, $(x_0, x_1, \ldots, x_{N-1})$, from the

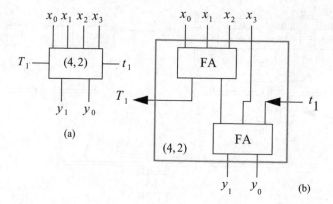

Figure 8.6   (a) The (4,2) compressor. (b) Full-adder implementation.

column and a number $l = N - 3$ of carry-in signals, $(t_1, t_2, \ldots, t_l)$, from the previous column. It outputs two signals, (y_1, y_0), in its column as well as l carry-out signals, $(T_1, T_2, \ldots, T_l)$, which are connected in the same order to the input lines of the next compressor to preserve the signal ordering. Additionally, the significance of an $(N, 2)$ compressor comes from the fact that horizontal as well as vertical propagation of the signals is allowed, instead of the traditional vertical direction of the full-adder tree.

In this section the problem of optimal design of compressors is addressed in two ways: first, the logic circuit which evaluates the PPST is improved using TGs obtaining an efficient logical redesign for the compressor basic building block, and second, such TGs are implemented in a very efficient way by using the νMOS principle. It is shown how the resulting νMOS basic compressors compare very favourably in power consumption and speed with previous conventional circuits.

8.4.2.1   The (4,2) compressor-design approach

A (4,2) compressor has five inputs of the same weight, $(x_0, x_1, x_2, x_3, t_1)$, and three outputs, (T_1, y_1, y_0), and it is generally implemented from two full adders with outputs y_1 and T_1 having the same weight (Fig. 8.6). Output T_1 is not a function of the input t_1, otherwise a ripple carry could occur.

The (4,2) compressor described here is based on TGs. Logic synthesis of an optimum threshold gate network for the (4,2) compressor has been carried out using integer linear programming. Figure 8.7 shows the corresponding threshold gate network which requires only two levels of TGs. The main characteristic of this network is the capability to obtain signals y_1 and y_0 with a delay which is the same and equal to the delay of two TGs.

8.4.2.2   Conventional via νMOS TG-based compressors – performance

This section illustrates the performance of both a (4,2) and a (6,2) compressor designed and laid out in a 0.8 μm double poly CMOS process. The correct operation under

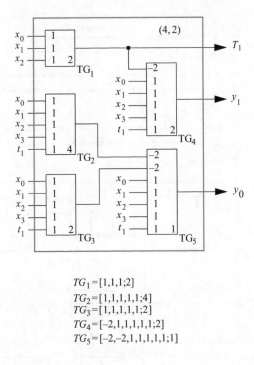

$$TG_1 = [1,1,1;2]$$
$$TG_2 = [1,1,1,1,1;4]$$
$$TG_3 = [1,1,1,1,1;2]$$
$$TG_4 = [-2,1,1,1,1,1;2]$$
$$TG_5 = [-2,-2,1,1,1,1;1]$$

Figure 8.7   TG implementation of a (4,2) compressor

process and ambient parameter variations is validated using HSPICE simulations of the extracted circuit that include Monte Carlo simulations and simulations using different standard worst case device parameters. Figure 8.8 shows simulated waveforms for the extracted (4,2) compressor for a sample of inputs. A new pattern is applied each 2.5 ns. The first five waveforms are the input signals and the next three ones are the output signals. For the purpose of comparison, a (4,2) compressor built from CMOS full adders is also designed and laid out. A well-known CMOS full adder topology with only 28 transistors is chosen as it requires a smaller circuit size and performs better in terms of speed and power dissipation than other CMOS implementations [251]. Sizing of transistors is carried out to optimise the transient performance of the full adder. The area for these two implementations is similar. Transient characteristics and average power are measured from post-layout simulation results using typical device parameters at a 5 V supply voltage. Table 8.1 compares the transient performance and the power-delay product of the two approximations for a (4,2) compressor. The results of a (6,2) compressor are also shown in Table 8.1. The delay corresponds to the worst case scenario; this is, situations where the inputs or the input sequence cause the slowest circuit operation. The power is measured using a random generated input sequence with 100 vectors. The table shows that a TG-based implementation is clearly advantageous over a conventional one. The operating frequency for both vMOS TG-based compressors is at least doubled when compared with previous conventional

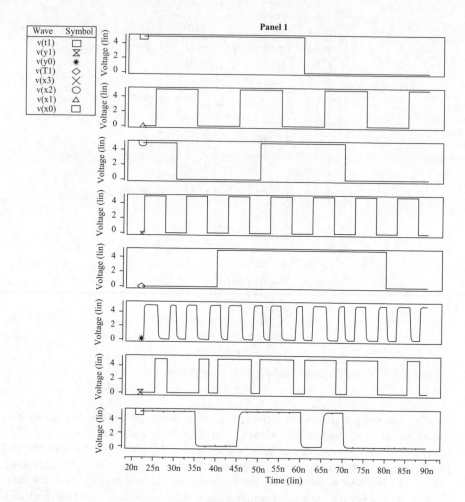

Figure 8.8   Simulated waveforms for a (4,2) compressor

approaches. Faster operation of the νMOS TG-based compressors does not increase the power dissipation too much. Power Delay Product (PDP) figures of the νMOS versions are also better. According to the obtained results, the νMOS designs are very efficient when compared with other reported implementations. They occupy a similar area, but exhibit a better time performance and power-delay product.

8.4.3   Sorting networks implemented as νMOS circuits

This section describes a hardware realisation of a binary sorting network (SN) based on the use of νMOS transistors. An n-input SN is a switching network with n outputs that generates an output which is a sorted (non-increasing order) permutation of inputs. Binary SNs are built from comparator cells which have two inputs and two

Table 8.1   Comparison of performance

Compressor		Conventional	TG-based
(4,2)	Delay time	3.1 ns	1.2 ns
	PDP @200 MHz	8.6 pJ	6.9 pJ
(6,2)	Delay time	4.5 ns	1.6 ns
	PDP @200 MHz	20.3 pJ	15.8 pJ

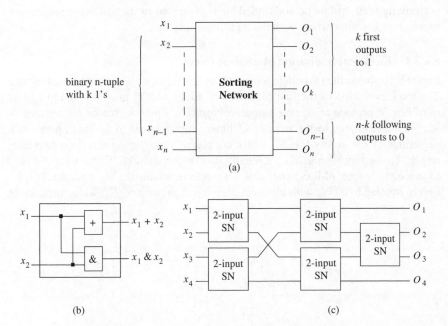

(a)

(b)   (c)

Figure 8.9   *(a) Sorting Network with k binary signal inputs equal to 1. (b) Logic gate implementation of the comparator cell (2-input SN). (c) Batcher's implementation of a 4-input SN*

outputs: one of them provides the maximum of both inputs and the other the minimum. The internal structure of the comparator depends on the application. The inputs can be either binary numbers, the comparator being then a complex element, or binary signals, in which case maximum and minimum become OR and AND operations respectively. Figure 8.9(a) shows the operation of an n-input sorter with k inputs equal to 1. Figure 8.9(b) shows the logic implementation of the comparator cell. The efficient implementation of an SN has for many years been an important subject of research [252]. The method proposed by Batcher was a milestone [253]. Figure 8.9(c) shows a 4-input SN implemented following Batcher's method.

A different approach to the implementation of a sorter will be next presented, which is based on the fact that each output of an n-input sorter depends only on the number of inputs equal to 1. The i-th output will be 1 if and only if at least i of the n inputs are 1. In this way, an n-input sorter can be seen as a cascaded two-block circuit. The first block provides an output which depends linearly on the number of 1s applied at the inputs. The second block takes this output signal and compares it with a set of n fixed values using a battery of comparators, thus providing the set of n output functions of an n-input sorter.

An electrical realisation of an 8-input sorter can be carried out following the above two-block approach and resorting to the νMOS transistor principle for its implementation, as it requires counting the number of 1s in the inputs, or equivalently performing their arithmetic addition. This is presented in the following subsection, together with a comparison with the traditional solution.

8.4.3.1 Electrical realisation of a sorter circuit

Figure 8.10 shows the two-stage schematic diagram of a νMOS based n-input sorter. The implementation of the first block resorts to the νMOS principle and to current mirroring to provide an analog output voltage, V_1, which increases proportionally, in a staircase shape, to the number of binary inputs equal to 1. This operation is performed by transistors M_1–M_4 which are biased in the strong inversion saturation region. Transistors M_2 and M_4 are equally sized n-channel νMOS transistors. M_1 and M_3 are equally sized pMOS transistors. The sorter inputs are the M_2 input gates capacitively coupled to its FG with identical coupling capacitances, C_u, which produce an

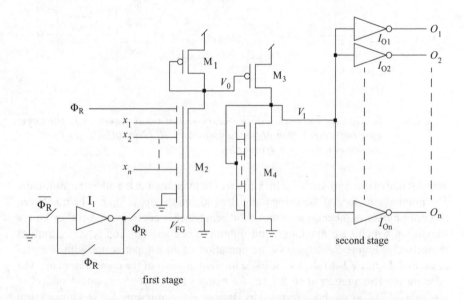

Figure 8.10   Two-stage schematic of a νMOS based n-input sorter

FG voltage, V_{FG}, linearly dependent on the sum of the inputs. However, with this circuit several input combinations with different numbers of 1s can give floating gate voltages below the threshold voltage of the n-channel MOS transistor which would not be properly distinguished. In order to avoid such a situation, the inverter I_1 as well as two additional inputs to transistor M_2 with coupling capacitances $C_u/2$ and C_o have been included. During the initialisation phase, $\Phi_R = 1$ switches that are on during this phase connect together M_2 FG as well as the output and input of I_1. Simultaneously the input terminals $x_1, x_2, \ldots, x_n$ are connected to ground (input switches not shown in Fig. 8.9). After initialisation, during processing mode, when $\Phi_R = 0$

$$V_{FG} = \left(\sum_{i=1}^{n} x_i \right) \cdot V_{DD} \cdot C_u/C_T + U_{inv(I_1)} - V_{DD} \cdot (C_u/2)/C_T \qquad (8.22)$$

where $U_{inv(I_1)}$ is the threshold of inverter I_1

$$C_T = (n + 1/2)C_u + C_{chan} + C_o \qquad (8.23)$$

Capacitance C_o is introduced by the extra grounded input in order to maintain M_2 saturated, even when the n inputs of the sorter are at logic '1' (see Section 3.3.5), and C_{chan} accounts for the transistor parasitic capacitances. V_{FG} controls the current through M_1 and M_3. Since M_4 is equal to M_2 the voltage produced at the M_4 drain terminal, $V_1 = V_{FG}$. With this scheme, the obtained analog output staircase shape voltage V_1 is robust to process parameter variations.

The second block comprises a set of comparators which are implemented as inverters. Each inverter is sized in such a way that its switching threshold falls between two given consecutive steps of the aforementioned staircase. For example, the output O_1 must be a logical '1' if there is at least an input at a logical '1' and so the switching threshold of inverter I_{O1} is fixed to $(V_1(0) + V_1(1))/2$, where $V_1(0)$ stands for the voltage at node V_1 when an all zero input vector is applied. $V_1(1)$ corresponds to the voltage at node V_1 when an input vector with only one '1' is applied.

Due to process parameter variations the voltages $V_1(i), i = 0, \ldots, n$, as well as the switching thresholds of the comparator-inverters can change from their nominal value. In order to reduce this sensitivity I_1 is made identical to I_{O1}. The role of capacitor $C_u/2$ becomes now clear; it assures that with all inputs at logical zero the V_1 voltage is under $U_{inv(I_{O1})}$, the switching threshold of inverter I_{O1}.

8.4.3.2 Design and evaluation of an 8-input sorter

This section illustrates the performance of an 8-input νMOS sorter designed and laid out in a 0.8 μm double poly CMOS process. In order to minimise excessive load of the comparators over the first block, the M_3-M_4 branch is replicated. Comparators are made of chains of inverters in order to regenerate the output signal to full logic swing. The operation is validated through HSPICE simulations of the extracted circuit including Monte Carlo simulations and simulations using different standard worst case device parameters. Figure 8.11 shows the simulation results for nodes V_0 and V_1 as functions of the number of inputs at logic '1'. It can be seen how the changes at V_0 are significantly reduced at V_1.

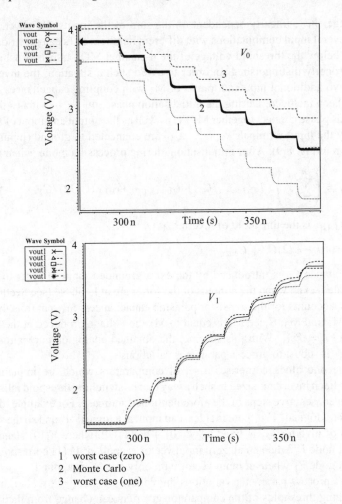

1   worst case (zero)
2   Monte Carlo
3   worst case (one)

Figure 8.11   Simulated behaviour of nodes V_0 and V_1 as functions of the number of inputs equal to 1 showing the stabilising action of the circuit

For the purpose of comparison, an 8-input sorter following the Batcher's conventional approach consisting of a network of comparator cells is also designed and laid out. There are 23 of such cells in an 8-input SN. Table 8.2 compares the area, the time performance and the power consumption of both sorters. Transient characteristics and average power are measured on post-layout simulation results using typical device parameters at a supply voltage of 5 V. The worst case delay time happens in situations when the inputs or the input sequence are such that the circuit operation is the slowest. In the conventional design an input vector exciting the true longest path has been used to measure that delay. In the νMOS counterpart an input vector consisting of only '1s' followed by an input vector consisting of only '0s' is employed. The power is measured using a random generated input sequence with 100 vectors.

Table 8.2   Area, time performance and power consumption of
vMOS and conventional sorters

	Area	Worst case delay	Power consumption (@175 MHz)
νMOS	5625 μm^2	4.1 ns	7.8 mW
Conventional	45400 μm^2	5.2 ns	8.2 mW

Compared with a conventional gate-based implementation, the νMOS design is very efficient in terms of area. It occupies an area that is nearly an order of magnitude smaller than in its conventional counterpart and, at the same time, it exhibits a better transient performance. Furthermore, the power consumption is found to be very dependent on the frequency in the conventional design as opposed to the νMOS design. For frequencies of 170 MHz and above, the νMOS sorter consumes less power than the conventional implementation.

Finally, it could be claimed that the transistor used in the sorter is not a real νMOS since the FG is not really floating (see Section 2.4.2). However, the FG is effectively floating for the range of operating frequencies of this circuit. It is true that the gate needs to be reset every few milliseconds; however, that should not be a problem as in that time the circuit is able to perform hundreds of thousands of operations.

8.4.3.3 νMOS sorters for arithmetic applications

The threshold functions produced at the outputs of the sorter circuit are involved in many arithmetic-like operations. This is illustrated below with two examples. The first example refers to the implementation of an (8 × 8) multiplier, while the second one describes a (15,4) counter which is used in the summation of the partial products in a parallel multiplier.

The (8 × 8) multiplier:

The (8 × 8) multiplier is based on the architecture shown in Fig. 8.12, originally proposed in [254]. The main component of this design, apart from the peripheral circuitry necessary for data scheduling, is a combinational functional block (F_Block) with 16 inputs and 9 outputs. Eight of the outputs correspond to threshold functions $T_2^{16}, T_4^{16}, T_6^{16}, T_8^{16}, T_{10}^{16}, T_{12}^{16}, T_{16}^{14}$ and T_{16}^{16}. The ninth is the parity function. The F_Block circuit is realised using a two-level network of capacitive TGs (17 gates). The F_Block described here uses the νMOS sorter circuit as the key component. Figure 8.13 shows the logic diagram. It consists only of a 16-input sorter, T^{16}, and a TG, T_9^{16}, realised using the ideas explained in the previous section. The output of the TG implements the parity function following the Muroga's method [255] as

$$\text{parity} = T_1^{16} - T_2^{16} + T_3^{16} - \cdots - T_{14}^{16} + T_{15}^{16} - T_{16}^{16} \tag{8.24}$$

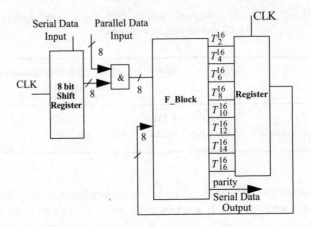

Figure 8.12 Serial/Parallel multiplier architecture

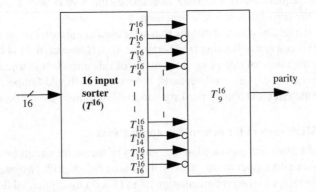

Figure 8.13 Logic diagram for the F_Block

The F_Block circuit using the νMOS sorter is designed and laid out in a 0.8 μm 5 V double poly CMOS process [214]. Figure 8.14 shows the simulated waveforms for the parity output of the extracted F_Block. The inputs correspond to a sequence of input patterns with an increasing number of ones: $(x_1, x_2, \ldots, x_{16}) = \{(0, 0, \ldots, 0), (0, 0, \ldots, 1), \ldots, (1, 1, \ldots, 1)\}$ starting at time $t = 60$ ns. A new pattern is applied each 7.5 ns. Clearly, the parity of the 16 input signals is correctly evaluated.

Transient characteristics and average power are measured on post-layout simulation results using typical device parameters at a supply voltage of 5 V. The power is measured using a randomly generated input sequence with 100 vectors. The worst case delay time is 4.5 ns and the power consumption is 13 mW at 100 MHz.

In order to validate the proposed circuit, its performance is compared with other solutions. Simulation results for the threshold-gate-based implementation of the

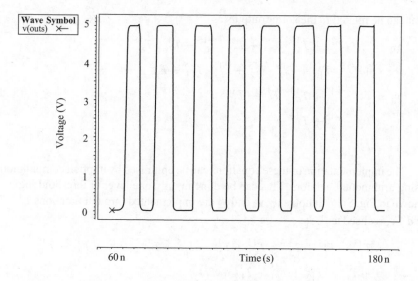

Figure 8.14   HSPICE simulation results for the parity output of F_Block

architecture in Fig. 8.12 [254] show a maximum clock frequency of around 30 MHz for the multiplier when implemented in a 1.2 μm technology. On the other hand, the multiplier based on a sorter circuit can work at frequencies in excess of 175 MHz. This is because the clock frequency of the former is mainly limited by the signal propagation through the F_Block.

Also, in order to compare these two approaches, the F_Block is designed and laid out following a conventional approach (NOR and NAND gates were used) using the same technological process. The worst case delay for this conventional design is over 11 ns and the power consumption at 66 MHz is 13 mW. Besides, it occupies an area between one and two orders of magnitude larger.

The (15,4) counter:

A second example illustrates a possible application of the sorter as a counter. This section describes the design methodology of a (15,4) counter.

A counter is a combinational circuit which outputs a binary number equal to the number of input lines that are at logical one. The summation of partial products in a parallel multiplier is traditionally performed using a full adder tree (full adders are a particular case of counters, the (3,2) counter). However, because the routing can be complicated, high-order counters are preferred. High-order counters are usually implemented using (3,2) counters because of the disadvantages of a direct implementation [250]. With the approach described a counter can be built directly from its logic equations. Let $(x_0, x_1, \ldots, x_{14})$ be the fifteen numbers to add in a (15,4) counter, and (y_3, y_2, y_1, y_0) be the counter output. Signals y_3, y_2, y_1 and y_0 are symmetric functions that can be related to the sorter outputs as

$$y_0 = T_1^{15} \cdot T_2^{15} + T_3^{15} \cdot T_4^{15} + T_5^{15} \cdot T_6^{15} + T_7^{15} \cdot T_8^{15}$$
$$+ T_9^{15} \cdot T_{10}^{15} + T_{11}^{15} \cdot T_{12}^{15} + T_{13}^{15} \cdot T_{14}^{15} + T_{15}^{15}$$
$$y_1 = T_2^{15} \cdot T_4^{15} + T_6^{15} \cdot T_8^{15} + T_{10}^{15} \cdot T_{12}^{15} + T_{14}^{15} \qquad (8.25)$$
$$y_2 = T_4^{15} \cdot T_8^{15} + T_{12}^{15}$$
$$y_3 = T_8^{15}$$

The implementation of these equations can be optimised by implicit computations, using arithmetic operators. This can be done in only one level of threshold logic, as shown in Fig. 8.15. Outputs y_0, y_1 and y_2 are implemented through functions T_8^{15}, T_4^7 and T_2^3, realised as TGs:

$$y_0 = T_1^{15} - T_2^{15} + T_3^{15} - T_4^{15} + T_5^{15} - T_6^{15} + T_7^{15} - T_8^{15} + T_9^{15} - T_{10}^{15}$$
$$+ T_{11}^{15} - T_{12}^{15} + T_{13}^{15} - T_{14}^{15} + T_{15}^{15}$$
$$y_1 = T_2^{15} - T_4^{15} + T_6^{15} - T_8^{15} + T_{10}^{15} - T_{12}^{15} + T_{14}^{15} \qquad (8.26)$$
$$y_2 = T_4^{15} - T_8^{15} + T_{12}^{15}$$
$$y_3 = T_8^{15}$$

Two (15,4) counters, one using the νMOS sorter circuit and νMOS TGs and another one following a conventional approach, are designed and laid out in the same technological process. Transient characteristics and average power are measured.

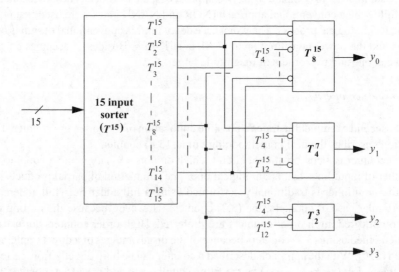

Figure 8.15   Logic diagram of the (15,4) counter

The worst case delay time for the νMOS solution is 8 ns and the power consumption is 12 mW at 66 MHz and it is very independent of the frequency. The conventional design has the same power consumption at 66 MHz, but the worst case delay is 11.25 ns.

8.4.4 A multi-input Muller C-element

A Muller C-element (where the C stands for concurrence) is a circuit widely used in the design of self-timing circuits to perform the function 'and' of events (transitions $1 \rightarrow 0$ or $0 \rightarrow 1$). The output of a Muller C-element is equal to the value of the input after all the inputs reach the same value; otherwise, it remains the same. It has been proven [235] that an m-input Muller C-element can be implemented using a single threshold gate with $(m + 1)$ inputs, and the simplest solution is obtained when the primary inputs have an associated weight of 1, the $(m + 1)$-th input (the feedback input) is affected by a weight of $(m - 1)$, and the threshold of the TG is m, as shown in Fig. 8.16(a).

The complexity of this logic element is high enough to use it as a demonstrator of a design approach based on νMOS-TG. On one hand, it shows that a complex functionality can be implemented by a single inverter. On the other hand, the existence of a feedback loop with a high associated weight $(m - 1)$ allows the testing of the signal regeneration capability that would guarantee the correct operation of threshold networks.

Figure 8.16(b) depicts the νMOS realisation of the logic diagram shown in Fig. 8.16(a), for $m = 8$. This circuit is designed and fabricated in a 5 V 0.8 μm double poly CMOS technology. Operation under process and ambient parameters was validated through extensive HSPICE simulations of the extracted circuit including Monte Carlo simulations and simulations using different standard worst case device parameters. Figure 8.17 shows the experimental responses. The waveform at

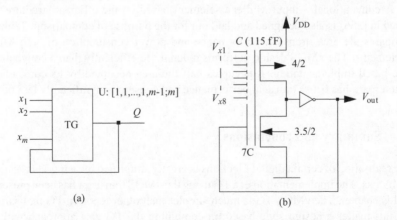

(a)   (b)

Figure 8.16 (a) *Threshold-gate-based m-input Muller C-element realisation;* (b) *Electrical diagram of a νMOS-based 8-input Muller C-element (transistor sizes and capacitance values are shown in the figure)*

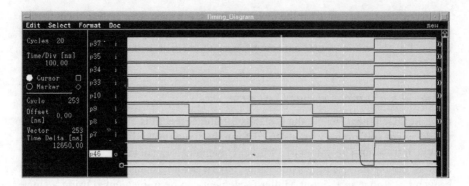

Figure 8.17   Experimental waveforms of the νMOS-based 8-input Muller C-element

Table 8.3   Performance parameters for νMOS and conventional 8-input Muller C-elements

	Area	Worst case delay	Power consumption (@100 MHz)
νMOS	4927 μm^2	1.8 ns	0.03 mW
Conventional	9942 μm^2	3.2 ns	0.54 mW

the bottom trace is the circuit output and the remaining waveforms correspond to the circuit input signals. Correct operation is obtained with a supply voltage down to 3 V.

A conventional 8-input Muller C-element based on the efficient structure proposed in [256] is also designed and laid out for the purpose of comparison. Table 8.3 compares the area, transient performance and power consumption of both Muller C-elements. The νMOS design performs much more efficiently than a conventional gate-based implementation. It occupies half the area occupied by its conventional counterpart, has better transient performance and consumes significantly less power.

8.5   Summary and conclusions

The realisation of certain digital functions is greatly simplified by using TGs as building blocks. The implementation of a TG using the νMOS transistor has been presented in this chapter. The νMOS TG is much simpler than other reported TG realisations. All this makes a design style based on combining the TG gate approach with the implementation of this building block using νMOS devices markedly interesting. The use of TGs as well as νMOS devices reduces the overall power × delay figure of merit of certain digital systems. This chapter has illustrated how to achieve this with several design examples. The reader can extend the illustrated design approach much further, to other digital systems that operate with even tighter design constraints in terms of the power consumption and voltage supply, and also adapt these circuits to new improved technological processes.

Notation

Δ	Delay of the final sum bits in an n-bit CLA (eq. (8.14)). Also delay difference between two realisations, traditional and hierarchical, of an n-bit adder and b-bit BCLA given by eq. (8.21)
Δ_{BCLA}	Circuit delay of a carry lookahead unit block (eq. (8.16))
Δ_{CLA}	Delay of a CLA block (eqs. (8.14), (8.16) and (8.20))
Δ_{PG}	Delay of an n-bit CLA due to carry generate/propagate unit generating P_i and G_i signals (eqs. (8.14) and (8.16))
Δ_{S}	Delay corresponding to the generation of S_i sum signals (eqs. (8.14) and (8.16))
Δ^{TG}	Delay of the n-bit adder in a hierarchical approach (eq. (8.20))
$\Delta_{\text{BCLA}}^{\text{TG}}$	Circuit delay of the BCLA unit based on TGs (eq. (8.20))
δ	Number of delay units per level (eq. (8.21))
Φ_{R}	Switch signal in the n-input sorter in Fig. 8.10 (has the value 1 during initialisation and the value 0 during the processing mode)
$A =$ $A_{n-1}, \ldots A_1, A_0$	Augend input to an n-bit CLA
$B =$ $B_{n-1}, \ldots B_1, B_0$	Addend input to an n-bit CLA
$\text{BCLA}(j, k)$	4-Bit BCLA unit corresponding to a generic j-th column and k-th row in an adder three implementation (Fig. 8.4 and Fig. 8.5)
C	Input capacitances (connected to V_{xi}) of the νMOS-based 8-input Muller C-element in Fig. 8.16(b)
C_{-1}	Carry input to the least significant position of an n-bit CLA
C_{chan}	Sum of M2 parasitic capacitances in the n-input sorter in Fig. 8.10
C_{Control}	Control input capacitance in Fig. 8.3
C_i	Input capacitance corresponding to the input x_i in Fig. 8.3 (eq. (8.2)). Also used in a different section to refer to the carry output of the i-th stage in an n-bit CLA obtained using G_i, P_i and C_{i-1} (see eq. (8.12))
C_{i-1}	Carry input to the i-th bit position of an n-bit CLA
C_{o}	M2 input capacitance connected to ground in the n-input sorter in Fig. 8.10

C_T	Total capacitance seen by the FG (eq. (8.2)). The same value has been assumed for all the devices unless the opposite is said
C_u	Input capacitances connecting input gates x_i of M2 (sorter inputs) and its FG (Fig. 8.10)
c	Extra control input of the νMOS TG in Fig. 8.3 used for the logic threshold adjustment
$F_k : F_{k-1} + F_{k-2}$	k-th Fibonacci's number
$f(x)$	Threshold function defined by $[w; T]$ where $w = [w_1, w_2, \ldots, w_n]$ (Fig. 8.2)
$f(x)$	Complement of a threshold function defined by $[-w; 1 - T]$ (Fig. 8.2)
$f(\bar{x}) : [-w; T - \sum w_i]$	Threshold function of complemented inputs (Fig. 8.2)
$\dot{f}(\bar{x}) : [w; 1 - T + \sum w_i]$	Complement of a threshold function of complemented inputs (Fig. 8.2)
G_i	Carry generate function (eq. (8.12))
O_i for $i = [1, n]$	Outputs of the sorting network
P_i	i-th Carry propagate function (eq. (8.12))
S_i	i-th Sum of the bits in an n-bit CLA (eq. (8.12))
T	TG threshold value (eq. (8.1))
T_1	Carry-out signal of a (4,2) compressor (Fig. 8.7)
T^{16}	16-Input sorter (Fig. 8.12)
T_i^{16} for $i = [2, 4, 6, 8, 10, 12, 14, 16]$	Outputs of the F_Block in Fig. 8.13 corresponding to the threshold functions
TG_i for $i = [1, 5]$	TGs used for implementing the (4,2) compressor (Fig. 8.7)
t_1	Carry-in signal to the (4,2) compressor (Fig. 8.7)
$U_{inv(I_1)}$	Switching threshold voltage of the inverter I_1 in Fig. 8.10
$U_{inv(I_{O1})}$	Switching threshold voltage of the inverter I_{O1} in Fig. 8.10
U_{inv}	Switching threshold voltage of the inverter (νMOS TG) in Fig. 8.3
V_c	Voltage applied to the control input c of the νMOS threshold gate in Fig. 8.3
V_{FG}	FG voltage at the FG of the νMOS TG in Fig. 8.3
$V_{xi} : x_i V_{DD}$	Voltage applied to the input x_i of the νMOS TG in Fig. 8.3 (eq. (8.2))
V_0	M2 drain voltage in the first block of the n-input sorter in Fig. 8.10
V_1	Analogue output voltage of the first block in the n-input sorter in Fig. 8.10
V_{out}	Output of the νMOS-based 8-input Muller C-element in Fig. 8.16(b)

V_{xi} for $i = [1,8]$ Inputs of the νMOS-based 8-input Muller C-element in
Fig. 8.16(b)

$W_T(C_j)$ Total weight for the TG implementing C_j given by eq. (8.19)

w_i i-th Input weights (eqs. (8.1) and (8.4))

x_i for $i = [1,n]$ Inputs

y_i i-th Output signals of the (4,2) compressor (Fig. 8.6). Also
counter outputs (Fig. 8.15)

Chapter 9
Summary and conclusions

This chapter summarises the most relevant results and conclusions that have been presented throughout the book.

Chapter 2 focused on the characterisation and modelling of the FGMOS transistor. From the analysis introduced in this chapter, the FGMOS transistor can be seen as a MOS transistor with the following characteristics:

(1) Multiple inputs (as many as there are input capacitors).
(2) Output resistance strongly dependent on the ratio between the capacitive coupling between the FG and the drain terminal and the total capacitance seen by the FG. The smaller this ratio, the larger the output resistance is.
(3) Reduced effective transconductance, which is scaled down by the effective input weight.

The first characteristic can be considered as an advantage with respect to the normal MOS transistor, whereas the reduction in the output resistance and transconductance may, in certain situations, represent drawbacks.

The second part of the chapter described some of the initial issues that designers encounter when starting to work with FGMOS devices, as well as solutions to them. The main two problems observed were as follows:

(1) The foundries do not provide models to simulate the FGMOS.
(2) There is an unknown amount of charge that can remain trapped at the FG during the fabrication process.

A very simple technique that does not require special device modelling to simulate the device was explained and compared with other existing methods that do need to model the transistor beforehand. Also, different techniques to control the charge at the FG were reviewed, together with a very simple layout technique, consisting of adding a contact to metal on the FG, which removes the initial trapped charge.

One modification to the FGMOS transistor for which the 'initial charge' is not a problem is the pseudo-floating-gate MOS device. This is an FGMOS device with a very high value resistor connected to the gate, filtering out low frequency signals. Since the trapped charge is equivalent to an initial DC condition, in this way it becomes irrelevant.

One of the main negative points of FGMOS transistors is the area required by the added input capacitors. Hence, ideally, these capacitors should be designed as small as possible. The problem is that the smaller these are, the higher the ratios between the parasitic capacitances and the total capacitance seen by the FG become. This may affect the performance of the circuits in different ways (as was shown in following chapters). Chapter 2 provided an analysis on how to choose the values of the input capacitances and an estimate of the overhead in terms of area.

Chapter 3 introduced basic circuit blocks that illustrated some of the interesting features of the FGMOS transistor. The most important ones are as follows:

(1)   The threshold voltage is controllable. This property can be used whenever the voltage requirements in a given design are too restrictive for a given nominal threshold voltage and a certain power supply voltage.

(2)   The transistor is a powerful computational element. It performs a weighted sum of its inputs in a high impedance node. The coefficients can be chosen by sizing the input capacitances appropriately.

(3)   The operating point at the gate of the transistor can be shifted as required by the circuit topology without the need for adding extra circuitry.

(4)   The transistor biased in the weak inversion saturation region can behave as a current controlled current source[79], which is a nonlinear function between the currents, with exponents dependent on the ratios between the input capacitances. In addition, the gain of the current source can be programmed by means of voltage sources connected to the transistor inputs.

The chapter also shows how to reduce the minimum required power supply voltage in a cascode current mirror with FGMOS transistors by using the voltage supply in one of the inputs to bias the devices and sizing the input capacitances accordingly.

Another interesting simple block is the FGMOS inverter in which an input can be used to program the switching threshold.

In an FGMOS comparator the versatility of the device was illustrated with four-input FGMOS devices in which each input was used for a different aim: effective input, offset cancellation, compensation of the degradation in output resistance caused by the gate to drain parasitic capacitance and reset.

The FGMOS transistor used in a switched current memory cell permits to program the cell in a wide range without the need of any other extra circuitry.

Finally some ideas for the implementation of very simple D/A converters making use of the summation property at the FG were also presented.

---

[79]   If its inputs are voltages generated by currents flowing through other transistors.

Chapters 4 and 5 demonstrated the potential of FGMOS devices biased in the strong inversion region for the design of low power and low voltage programmable analog G_m-C filters. The designs presented showed how:

(1) The use of the FGMOS transistor simplified the designs enormously due to the addition of the input voltages at the FG.
(2) An immediate consequence of (1) is that fewer devices are needed, which means fewer internal nodes. This is advantageous from the point of view of frequency response.
(3) The noise is lower as the circuit topologies get simpler.
(4) The voltage supply can be scaled down by optimising the voltage ranges at the devices. This can be achieved by shifting down the effective threshold voltage with the help of one of the transistor's inputs. The input range is also increased this way, due to the attenuation of the effective input voltage at the FG.
(5) The power consumption decreases as a consequence of the lower power supply voltage and the smaller number of active devices.
(6) If the transistor operates in the strong inversion ohmic region the Total Harmonic Distortion is lower than in a MOS transistor with equivalent voltage swings at the gate. The reason for this is that the C_{GD} term appears subtracted in the quadratic V_{DS} term, which reduces the nonlinear contribution to the current function.

There are also some drawbacks in using the device, which can be summarised as follows:

(1) As was mentioned before, the output resistance is reduced because of the capacitive coupling between the FG and the drain terminal. A possible way to deal with this if necessary is to add cascode transistors.
(2) If the effective threshold voltage is scaled down by using extra inputs connected to the voltage supplies the PSRR degrades. It also becomes dependent on the mismatch between the capacitances.
(3) The CMRR is affected by the mismatch between the input capacitances.
(4) The transconductance and the frequency range are lower than the transconductance and frequency range that could be obtained with equally sized MOS transistors driving the same values of currents. However, whether this causes a problem or not depends on the application. Also, in general, assumption of having a similar circuit only with MOS transistors is not realistic since using the FGMOS transistor only makes sense whenever normal MOS cannot be used to obtain the required performance in terms of voltage, power and signal ranges.
(5) Although a double poly technology is not strictly required, if low quality capacitors are used instead of the FGMOS the circuit performance will degrade.

Chapters 6 and 7 explored the design of low power and low voltage G_m-C and log-domain filters, taking advantage of the functionality of the FGMOS transistor biased in the weak inversion saturation region and using the Translinear Principle, which was reformulated for the FGMOS. The most relevant conclusions extracted from these chapters were as follows:

(1) Translinear loops created with FGMOS devices simplify the circuit topologies. This brings with it a reduction in the number of internal nodes, which improves the frequency response, and also lowers noise and power consumption.
(2) The bulk and source terminal do not need to be used to implement the translinear equation. Because of this they can be connected to constant voltages, which reduces the risk of instabilities.
(3) There is no need of stacking transistors to implement the translinear equation, which is beneficial for low power supply voltage operation.
(4) There are many possible alternatives when programming the design, thanks to the controllability property of the FGMOS device.

The main disadvantage of using FGMOS in weak inversion are as follows:

(a) The mismatch between input capacitances together with the parasitic coupling between the drain and the FG affects the Total Harmonic Distortion.
(b) In log-domain implementations, the signal is not as compressed in the FG as when MOS transistors are used instead because of the smaller than one capacitive input weight.

Finally, Chapter 8 moved away from analog design, illustrating the advantages of using FGMOS devices in digital circuits. The chapter showed how logic circuit design based on Threshold Gates implemented with FGMOS transistors is an alternative to traditional logic gate design. Using threshold gates as main building blocks in a system realisation enables a reduction in both the number of gates and the gate levels, with the consequent saving in power consumption, mainly for high-frequency operation.

So, where do we go from here? Although this book has a chapter on digital design and reviews the main issues related to power and voltage reduction in digital circuits, it is mainly a text on analog design. As many have admitted before, analog design is an art based on the combination of the different degrees of freedom in the transistors' behavioural equations. If properly combined, the result is a beautiful piece that meets the consumers' specifications. The FGMOS transistor is like a wonderful new colour palette, a transistor with added degrees of freedom. This opens up a whole new world of possibilities for the designer. The specifications are the same, but the limitations on how to achieve them are fewer. This book has shown just a tiny sample of topologies that can be created by playing with the extra degrees of freedom, achieving a far superior performance than the one that could have been obtained with normal MOS devices, mostly in terms of voltage ranges and power. There are still many other circuit topologies to invent and explore in the context of low power and low voltage electronics. Also, due to the problem of the trapped charge at the FG (which has only recently been solved), the device has not been widely used. As a consequence, not many electronic systems had been reported that made use of the device in a context other than memories. It is now time to start exploiting the full potential of the transistor in more complex systems.

A very interesting property of the FGMOS transistor is that theoretically each device can be independently controlled. This could be a very promising possibility in certain applications that require a high level of programmability, tunability or calibration. An example could be making robust the circuit drivers in polysilicon technologies for TFT (Thin Film Transistors) displays. The main problem of this kind of technologies is related to the large variations among the devices. A calibration mechanism could be devised to adjust the drivers by changing the bias voltages in one of the inputs of TFT devices with floating gates. Another example is low power Analog Field Programmable Arrays (FPAA), in which different analog blocks could be programmed by changing the bias voltages in certain FGMOS devices.

In the digital arena, many new topologies, different from the ones presented here, could be invented to take advantage of compact threshold gate topologies built with the FGMOS device, based on the threshold decision principle.

A point worth stressing is that the basic property of the FGMOS transistor, its floating gate, may also be its most important limitation in any future technologies in which a leaky gate oxide might render the transistor useless for certain applications. In this case, the pseudo-FGMOS described in Chapter 2 could be used bearing in mind the limitations it has with respect to the normal FGMOS, as was previously explained.

Ultimately, there are plenty of electronic systems that could benefit from exploiting the capabilities of the FGMOS device and it is left to the creative mind of the circuit designer to decide how to combine all the degrees of freedom and come up with an optimum design. Having more variables to play with might make the process of finding the solution more difficult; nevertheless, the benefits make the challenge worthwhile in the end.

References

1 Vittoz, E. A.: 'Analog VLSI signal processing: why, where, and how?', *Analog Integrated Circuits and Signal Processing*, 1994 **8**, pp. 27–44

2 Luryi, S., Xu, J., and Zaslavski, A., (eds.): *Future trends in microelectronics. The road ahead* (Wiley-Interscience, 1999) ISBN: 0 471 32183 4

3 Lyon, R. F., and Mead, C.: 'An analog electronic cochlea', *IEEE Transactions on Acoustics, Speech and Signal Processing*, 1988 **36**(7), pp. 1119–34

4 Stotts, L. J.: 'Introduction to implantable biomedical IC design', *IEEE Circuits Devices Mag.*, 1989 **5** (1), pp. 12–18

5 Von Kaenel, V., Macken, P., and Degrauwe, M. G. R.: 'A voltage reduction technique for battery operated systems', *IEEE Journal of Solid-State Circuits*, 1990 **25**, pp. 1136–40

6 Yúfera, A., Rueda, A., Munoz, J. M., Doldan, R., Léger, G., and Rodríguez Villegas, E. O.: 'A tissue impedance measurement chip for myocardial ischemia detection', *IEEE Transactions on Circuits and Sytems I: Fundamental Theory and Applications*', 2005 **52**(12), pp. 2620–8

7 Baltes, H., and Brand, O. (invited keynote): 'CMOS integrated microsystems and nanosystems'. Proceedings of the *SPIE Conference on Smart Electronics and MEMS*, 1999 **3673**, pp. 2–10

8 'The National Technology Roadmap for Semiconductors. Technology Needs', Technical report, Semiconductor Industry Association, 2000.

9 El-Adawy, A. A., and Soliman, A. M.: 'A low-voltage single input class AB transconductor with rail-to-rail input range', *IEEE Transactions on Circuits and Systems-I*, 2000 **47**, pp. 236–42

10 Duque-Carrillo, J. F., Ausín, J. L., Torelli, G., Valverde, J. M., and Domínguez, M. A.: '1V rail-to-rail operational amplifiers in standard CMOS technology', *IEEE Journal of Solid State Circuits*, 2000 **35**, pp. 33–44

11 Lin, C., and Ismail, M.: 'Robust design of LV/LP low-distortion CMOS rail-to-rail input stages', *Analog Integrated Circuits and Signal Processing*, 1999 **21**, pp. 153–61

12 Ferri, G., Sansen, W., and Peluso, V.: 'A low-voltage fully-differential constant-G_m rail-to-rail CMOS operational amplifier', *Analog Integrated Circuits and Signal Processing*, 1998 **16**, pp. 5–15

13 deLangen, K. J., and Huijsing, J. H.: 'Compact low-voltage power-efficient operational amplifier cells for VLSI', *IEEE Journal of Solid State Circuits*, 1998 **33**, pp. 1482–96

14 Gregorian, R., and Temes, G. C.: *Analog MOS integrated circuits for signal processing* (Wiley-Interscience, 1986) ISBN: 0 471 09797 7

15 Setty, S., and Toumazou, C.: 'N-folded cascode technique for high frequency operation of low voltage opamps', *IEE Electronics Letters*, 1996 **32**, pp. 955–7

16 Xu, G., Embabi, S. H. K., Hao, P., and Sanchez-Sinencio, E.: 'A low voltage fully differential nested G_m capacitance compensation amplifier: analysis and design', Proceedings of the *IEEE International Symposium on Circuits and Systems*, 1999 **2**, pp. 606–9

17 Wu, J., and Chang, K.: 'MOS charge pumps for low-voltage operation', *IEEE Journal of Solid State Circuits*, 1998 **33**, pp. 592–7

18 Duister, T. A. F., and Dijkmans, E. C.: 'A -90dB THD rail-to-rail input opamp using a new local charge pump in CMOS', *IEEE Journal of Solid State Circuits*, 1998 **33**, pp. 947–55

19 St. Pierre, R.: 'Low-power BiCMOS opamp with integrated current mode charge pump'. Proceedings of the *IEEE European Solid-State Circuits Conference*, 1999 pp. 70–3

20 Shin, J., Chung, I., Park, Y., and Min, H.: 'A new charge pump without degradation in threshold voltage due to body effect', *IEEE Journal of Solid State Circuits*, 2000 **35**, pp. 1227–30

21 Tsividis, Y.: 'Externally linear, time-invariant systems and their application to companding signal processors', *IEEE Transactions on Circuits and Systems-II*, 1997 **44**, pp. 65–85

22 Frey, D., Tsividis, Y. P., Efthivoulidis, G., and Krishnapura, N.: 'Syllabic-companding log domain filters', *IEEE Transactions on Circuits and Systems-II: Analog and Digital Signal Processing*, 2001 **48** (4), pp. 329–39

23 Mulder, J., Serdijn, W. A., van der Woerd, A. C., and van Roermund, A. H. M.: 'An instantaneous and syllabic companding translinear filter', *IEEE Transactions on Circuits and Systems-II*, 1998 **45**, pp. 150–54

24 Tsividis, Y.: 'Minimizing power dissipation in analogue signal processors through syllabic companding', *IEE Electronics Letters*, 1999 **35**, pp. 1805–7

25 Khoury, J. M.: 'On the design of constant settling time AGC circuits', *IEEE Transactions on Circuits and Systems-II*, 1998 **45**, pp. 283–94

26 Toumazou, C., Ngarmnil, J., and Lande, T. S.: 'Micropower log-domain filter for electronic cochlea', *IEE Electronic Letters*, 1994 **30**(22), pp. 1839–41

27 El-Masry, E. I., and Wu, J.: 'CMOS micropower universal log-domain biquad', *IEEE Transactions on Circuits and Systems I*, 1999 **46**, pp. 389–92

28 Fragnière, E., and Vittoz, E.: 'A log-domain CMOS transcapacitor: design, analysis and applications', *Analog Integrated Circuits and Signal Processing*, 2000 **22**, pp. 195–208

29 Fox, R. M., and Nagarajan, M.: 'Multiple operating points in a CMOS log-domain filter', *IEEE Transactions on Circuits and Systems-II: Analog and Digital Signal Processing*, 1999 **46**(6), pp. 705–10

30 Degrauwe, M. G., Rijmenents, J., Vittoz, E. A., and DeMan, H. J.: 'Adaptive biasing CMOS amplifiers', *IEEE Journal of Solid State Circuits*, 1982 **17**, pp. 522–8

31 Klinke, R., Hosticka, B. J., and Pfleiderer, H. J.: 'A very-high-slew-rate CMOS operational amplifier', *IEEE Journal of Solid State Circuits*, 1989 **24**, pp. 744–6

32 Vittoz, E., and Fellrath, J.: 'CMOS analog integrated circuits based on weak inversion operation', *IEEE Journal of Solid State Circuits*, 1977 **12**, pp. 224–31

33 Horowitz, M., Indermaur, T., and Gonzalez, R.: 'Low-power digital design'. Proceedings of the *IEEE Symposium on Low Power Electronics*, 1994 pp. 8–11

34 Chandrakasan, A., Sheng, S., and Brodersen, R. W.: 'Low-power CMOS digital design', *IEEE Journal of Solid-State Circuits*, 1992 **27**, pp. 473–84

35 Indermaur, T., and Horowitz, M.: 'Evaluation of charge recovery circuits and adiabatic switching for low power CMOS design', Symposium on *Low-Power Electronics*, 1994 pp. 102–3

36 Hasler, P., and Lande, T. S.: 'Overview on floating-gate devices, circuits, and systems', *IEEE Transactions on Circuits and Systems II: Analog and Digital signal Processing*, 2001 **48**

37 Kahng, D., and Sze, S. M.: 'A floating-gate and its application to memory devices', *Bell Syst. Tech. J.*, 1967 **46**(4), pp. 1288–95

38 Holler, M., Tam, S., Castro, H., and Benson, R.: 'An electrically trainable artificial neural network with 10240 'floating gate' synapses', Proceedings of the *IEEE International Joint Conference on Neural Networks*, Vol. II, 1989, pp. 191–6

39 Lai, S.: 'Flash memories: where we were and where we are going', Proceedings of the *IEEE International Electron Devices Meeting*, 1998, pp. 971–4

40 Sackinger, E., and Guggenbuhl, W.: 'An analog trimming circuit based on a floating-gate device', *IEEE Journal of Solid-State Circuits*, 1988 **23**(6), pp. 1437–40

41 Yu, C. G., and Geiger R. L.: 'Very low voltage operational amplifiers using floating gate Mos transistor'. Proceedings of the *IEEE International Symposium on Circuits and Systems*, Vol. II, May 3–6 1993, pp. 1152–55

42 Lenzlinger, M., and Snow, E. H.: 'Fowler-Nordheim tunneling into thermally grown SiO_2', *Journal of Applied Physics*, 1969 **40**(1), pp. 278–83

43 Thomsen, A., and Brooke, M. A.: 'A floating gate MOSFET with tunneling injector fabricated using a standard double-polysilicon CMOS process', *IEEE Electronics Device Letters*, 1991 **12**(3), pp. 111–13

44 Op't Eynde, F., and Zorio, C.: 'An EEPROM in a standard CMOS technology'. Proceedings of the *23rd European Solid-State Circuits Conference*, Southampton, UK, Sep.–Oct. 1997

45 Lanzoni, M., Briozzo, L., and Riccò, B.: 'A novel approach to controlled programming of tunnel-based floating-gate MOSFET's', *IEEE Journal of Solid-State Circuits*, 1994 **29**(2), pp. 147–50

46 Gao, W., and Snelgrove, W. M.: 'Floating gate charge-sharing: a novel circuit for analog trimming'. Proceedings of the *IEEE International Symposium on Circuits and Systems*, 1994, pp. 315–18

47 Thomsen, A., and Brooke, M. A.: 'Low control voltage programming of floating gate MOSFETs and applications', *IEEE Transactions on Circuits and Systems, I: Fundamental Theory and Applications*, 1994 **41**(6), pp. 443–52

48 Lanzoni, M., Tondi, G., Galbiati, P., and Ricco, B.: 'Automatic and continuous offset compensation of MOS operational amplifiers using floating-gate transistors', *IEEE Journal of Solid-state Circuits*, 1998 **33**, pp. 287–90

49 Jackson, S. A., Killens, J. C., and Blalock, B. J.: 'A programmable current mirror for analog trimming using single-poly floating-gate devices in standard CMOS technology', *IEEE Transactions on Circuits and Systems II: Analog and Digital Signal Processing*, 2001 **48**, pp. 100–6

50 Shibata, T., and Ohmi, T.: 'A functional MOS transistor featuring gate-level weighted sum and threshold operations', *IEEE Transactions on Electron Devices*, 1992 **39**(6), pp. 1444–55

51 Shibata, T., and Ohmi, T.: 'Neuron MOS binary-logic integrated circuits- Part I: design fundamentals and soft-hardware-logic circuit implementations', *IEEE Transactions on Electron Devices*, 1993 **40**(3), pp. 570–6

52 Shibata, T., and Ohmi, T.: 'Neuron MOS binary-logic integrated circuits- Part II: simplifying techniques of circuit configuration and their practical applications', *IEEE Transactions on Electron Devices*, 1993 **40**(5), pp. 974–9

53 Shibata, T., Kotani, K., and Ohmi, T.: 'Real-time reconfigurable logic circuits using Neuron MOS transistors', *ISSCC Tech. Dif.*, 1993, pp. 238–9

54 Hany, T., Teranishi, K., and Kameyama, M.: 'Multiple-valued logic-in-memory VLSI based on a floating-gate-MOS pass-transistor network', Proceedings of the *IEEE International Solid-State Circuits Conference*, 1998, pp. 194–5

55 Kaeriyama, S., Hanyu, T., and Kameyama, M.: 'Arithmetic-oriented multiple-valued logic-in-memory VLSI based on current-mode logic', Proceedings of the *30th IEEE International Symposium on Multiple-Valued Logic*, 2000, pp. 438–43

56 Hanyu, T.: 'Challenge of a multiple-valued technology in recent deep-submicron VLSI'. Proceedings of the *31st IEEE International Symposium on Multiple-Valued Logic*, 2001, pp. 241–4

57 Hanyu, T., Kameyama, M., Shimabukuro, K., and Zukeran, C.: 'Multiple-valued mask-programmable logic array using one-transistor universal-literal circuits', Proceedings of the *31st IEEE International Symposium on Multiple-Valued Logic*, 2001, pp. 167–72

58 Weber, W., Prange, S. J., Thewes, R., Wohlrab, E., and Luck, A.: 'On the application of the neuron MOS transistor principle for modern VLSI design', *IEEE Transactions on Electron Devices*, 1996 **43**(10)

59 Fujita, O., and Amamiya, Y.: 'A floating-gate analog memory device for neural networks', *IEEE Transactions on Electron Devices*, 1993 **40**(11), pp. 2029–35

60 Shima, T., and Rinnert, S.: 'Multiple-valued memory using floating gate devices', *IEICE Transaction on Electronics*, 1993 **E76-C**(3), pp. 393–402

61 Kosaka, H., Shibata, T., Ishii, I., and Ohmi, T.: 'An excellent weight-updating-linearity EEPROM synapse memory cell for self-learning neuron-MOS neural networks', *IEEE Transactions on Electron Devices*, 1995 **42**(1), pp. 135–43

62 Shibata, T., Kosaka, H., Ishii, H., and Ohmi, T.: 'A neuron-MOS neural network using self-learning-compatible synapse circuits', *IEEE Journal of Solid-State Circuits*, 1995 **30**(8), pp. 913–22

63 Card, H. C., and Moore, W. R.: 'EEPROM synapses exhibiting pseudo Hebbian plasticity', *IEE Electronic Letters*, 1989 **25**, pp. 805–6

64 Lee, B. W., Sheu, B. J., and Yang, H.: 'Analog floating gate-synapses for general purpose VLSI neural computation', *IEEE Transactions on Circuits Systems*, 1991 **38**, pp. 654–7

65 Durffee, C. A., and Shoucair, F. S.: 'Comparison of floating gate neural network memory cells in standard VLSI CMOS technology', *IEEE Transactions on Neural Networks*, 1992 **3**(3), pp. 347–53

66 Hasler, P., Diorio, C., Minch, B. A., and Mead, C.: 'Single transistor learning synapse with long term storage', Proceedings of the *IEEE International Symposium on Circuits and Systems*, 1995, pp. 1660–3

67 Diorio, C., Hasler, P., Minch, B. A., and Mead, C.: 'A single-transistor silicon synapse', *IEEE Transactions on Electron Devices*, 1996 **43**(11), pp. 1972–80

68 Hasler, P., Minch, B. A., Diorio, C., and Mead, C.: 'An autozeroing amplifier using pFET hot-electron injection', Proceedings of the *IEEE International Symposium on Circuits and Systems*, 1996 **3**, pp. 325–9

69 Diorio, C., Hasler, P., Minch, B. A., and Mead, C.: 'A complementary pair of four-terminal silicon synapses', *Analog Integrated Circuits and Signal Processing*, 1997 **13**, pp. 153–66

70 Diorio, C., Hasler, P., Minch, B. A., and Mead, C. A.: 'Floating-gate MOS learning array with locally computed weight updates', *IEEE Transactions on Electronics Devices*, 1997 **44**, pp. 2281–9

71 Hasler, P., Minch, B. A., and Diorio, C.: 'An autozeroing floating-gate band-pass filter'. Proceedings of the *IEEE International Symposium on Circuits and Systems*, 1998

72 Hasler, P., Diorio, C., and Minch, B. A.: 'Continuous-time feedback in floating-gate MOS circuits'. Proceedings of the *IEEE International Symposium on Circuits and Systems*, Vol. 3, Monterey, CA, 1998

73 Apsel, A., Stanford, T., and Hasler, P.: 'An adaptive front end for olfaction'. Proceedings of the *IEEE International Symposium on Circuits and Systems*, Vol. 3, Monterey, CA, 1998

74 Hasler, P., and Dugger, J.: 'Correlation learning rule in floating-gate pFET synapses'. Proceedings of the *IEEE International Symposium on Circuits and Systems*, 1999, pp. 387–90

75 Hasler, P., Minch, B. A., and Diorio, C.: 'Adaptive circuits using pFET floating-gate devices'. Proceedings of the *Advanced Research VLSI Conference*, 1999, pp. 215–31

76 Hasler, P., Kucic, M., and Minch, B. A.: 'A transistor-only circuit model of the autozeroing floating-gate amplifier'. Proceedings of the *IEEE Midwest Conference on Circuits and Systems*, 1999, pp. 157–60

77 Hasler, P.: 'Continuous-time feedback in floating-gate MOS circuits', *IEEE Transactions on Circuits and Systems II: Analog and Digital Signal Processing*, 2001 **48**, pp. 56–64

78 Hasler, P., Minch, B. A., and Diorio, C.: 'An autozeroing floating-gate amplifier', *IEEE Transactions on Circuits and Systems II: Analog and Digital Signal Processing*, 2001 **48**, pp. 74–82

79 Cohen, M., and Cauwenberghs, G.: 'Floating-gate adaptation for focal-plane online nonuniformity correction', *IEEE Transactions on Circuits and Systems II: Analog and Digital Signal Processing*, 2001 **48**, pp. 83–9

80 Kucic, M., Low, A., Hasler, P., and Neff, J.: 'A programmable continuous-time floating-gate Fourier processor', *IEEE Transactions on Circuits and Systems II: Analog and Digital Signal Processing*, 2001 **48**, pp. 90–99

81 Hasler, P., Stanford, T., and Minch, B.: 'A second-order section built from autozeroing floating-gate amplifiers', *IEEE Transactions on Circuits and Systems II: Analog and Digital Signal Processing*, 2001 **48**, pp. 116–20

82 Yang, K., and Andreou, A. G.: 'A multiple input differential amplifier based on charge sharing on a floating-gate MOSFET', *Analog Integrated Circuits and Signal Processing*, 1994 **6**, pp. 197–208

83 Glasser, L. A.: 'A UV write-enabled PROM', *Chapel Hill Conference on VLSI*, Computer Science Press, (Rockville, MD, 1985) pp. 61–5

84 Kerns, D. A., Tanner, J., Sivilotti, M., and Luo, J.: 'CMOS UV-writable nonvolatile analog storage', Sequin, C. H. (ed.) *Advanced Research in VLSI*, MIT Press (Cambridge, MA, 1991) pp. 245–61

85 Benson, R. G., and Kerns, D. A.: 'UV-activated conductances allow for multiple time scale learning', *IEEE Transactions on Neural Networks*, 1993 **4**(3), pp. 434–40

86 Yang, K., and Andreou, A. G.: 'Subthreshold analysis of floating-gate MOSFET's'. Proceedings of the *10th Biennial University Government Industry Microelectronics Symposium*, 1993, pp. 141–4

87 Hasler, P., Minch, B., and Diorio, C.: 'Floating-gate devices: they are not just for digital memories anymore', Proceedings of the *IEEE International Symposium on Circuits and Systems*, 1999 **2**, pp. 388–91

88 Minch, B. A., Diorio, C., Hasler, P., and Mead, C. A.: 'Translinear circuits using subthreshold floating-gate MOS transistors', *Analog Integrated Circuits and Signal Processing*, 1996 **9**(2), pp. 167–79

89 Minch, B. A.: 'Analysis, Synthesis, and Implementation of Networks of Multiple-Input Translinear Elements', Ph.D. Thesis, Caltech, Pasadena, CA, 1997

90 Harrison, R., Hasler, P., and Minch, B.: 'Floating-gate CMOS analog memory cell array'. Proceedings of the *IEEE International Symposium on Circuits and Systems*, 1998, pp. 204–7

91 Harrison, R., Bragg, J. A., Hasler, P., Minch, B. A., and Deweerth, S. P.: 'A CMOS programmable analog memory-cell array using floating-gate circuits', *IEEE Transactions on Circuits and Systems II: Analog and Digital Signal Processing*, 2001 **48**, pp. 4–11

92 Hasler, P.: 'Foundations of learning in analog VLSI', Ph.D. Dissertation, California Institute of Technology, Pasadena, CA, 1997

93 Hasler, P., Andreou, A. G., Diorio, C., Minch, B. A., and Mead, C. A.: 'Impact ionization and hot-electron injection derived consistently from Boltzmann transport', *VLSI Design*, 1998 **8**(1–4), pp. 455–61

94 Ramírez-Angulo, J., Choi, S. C., and Gonzalez-Altamirano, G.: 'Low-voltage OTA architectures using multiple input floating gate transistors', *IEEE Transactions on Circuits and Systems*, 1995 **42**(12), pp. 971–4

95 Ramírez-Angulo, J., Choi, S. C., and González Altamirano, G.: 'Ultracompact lo-voltage analog CMOS multiplier using multiple-input floating gate transistors', Proceedings of the *European Solid-State Circuits Conference*, 1996, pp. 99–103

96 Ramírez-Angulo, J.: 'A +−0.75 V BiCMOS quadrant analog multiplier with rail-rail input signal swing', Proceedings of the *IEEE International Symposium on Circuits and Systems*, Atlanta, Georgia, 1996, pp. 242–5

97 Reza Mehrvarz, H., and Yee Kwok, C.: 'A novel multi-input floating-gate MOS four-quadrant analog multiplier', *IEEE Journal of Solid-State Circuits*, 1996 **31**(8), pp. 1123–31

98 Chen, J-J., Liu, S-I., and Hwuang, Y-S.: 'Low-voltage single supply four-quadrant multiplier using floating-gate MOSFETs', Proceedings of the *IEEE International Symposium on Circuits and Systems*, June 1997, pp. 237–48

99 Wong, L. S., Kwok, C. Y., and Rigby, G. A.: '1-V CMOS D/A converter with a multi-input floating-gate MOSFET', *IEEE Journal of Solid-State Circuits*, 1999 **34**(10), pp. 1386–90

100 Inoue, T., Nakane, H., Fukuju, Y., and Sánchez-Sinencio, E.: 'A low-voltage fully-differential current-mode analog CMOS integrator using floating-gate MOSFETS', *IEEE International Symposium on Circuits and Systems*, Vol. IV, Geneva, June 2000, pp. 145–8

101 Ramírez-Angulo, J., and López, A. J.: 'MITE circuits: the continuous-time counterpart to switched-capacitor circuits', *IEEE Transactions on Circuits and Systems II: Analog and Digital Signal Processing*, 2001 **48**, pp. 45–55

102 Muñoz, F., Torralba, A., Carvajal, R. G., Tombs, J., and Ramírez-Angulo, J.: 'Floating-gate-based tunable CMOS low-voltage linear transconductor and its application to HF GM-C filter design', *IEEE Transactions on Circuits and Systems II: Analog and Digital Signal Processing*, 2001 **48**, pp. 106–10

103 Ramírez-Angulo, J., Carvajal, R. G., Tombs, J., and Torralba, A.: 'Low-voltage CMOS op-amp with rail-to-rail input and output signal swing for continuous-time signal processing using multiple-input floating-gate transistors', *IEEE Transactions on Circuits and Systems II: Analog and Digital Signal Processing*, 2001 **48**, pp. 111–16

104 Ramírez-Angulo, J., Quintana, V., and Choi, S. C.: 'Analog VLSI fuzzy processor architectures based on multiple-input floating gate transistors', Proceedings of the *World Automation Congress*, 1996, pp. 539–44

105 Ramírez-Angulo, J., Quintana, V., Choi, S. C., Zrilic, D., and de Luca, A.: 'Charge mode defuzzifiers using multiple input floating-gate transistors',

Proceedings of the *International Journal of Intelligent Automation and Soft Control, Special Issue on Autonomous Control Engineering*, 1997 **3**(1), pp. 5–12

106 Yu, N. M., Shibata, T., and Ohmi, T.: 'A real-time center-of mass tracker circuit implemented by neuron MOS technology', *IEEE Transactions on Circuits and Systems II: Analog and Digital Signal Processing*, 1998 **45**(4), pp. 495–503

107 Berg, Y., and Lande, T. S.: 'Programmable floating-gate MOS logic for low-power operation', Proceedings of the *IEEE International Symposium on Circuits and Systems*, Hong-Kong, 1997, pp. 1792–5

108 Lande, T. S., and Berg, Y.: 'Ultra low-voltage transconductance amplifier', Proceedings of the *IEEE International Conference on Electronics, Circuits and Systems (ICECS)*, 1997 **1**, pp. 333–6

109 Berg, Y., and Lande, T. S.: 'Ultra low-voltage current mirrors and pseudo differential pairs', Proceedings of the *Eleventh Annual IEEE International ASIC Conference*, 1998, pp. 109–14

110 Berg, Y., and Lande, T. S.: 'Area efficient circuit tuning with floating-gate techniques', Proceedings of the *IEEE International Symposium on Circuits and Systems*, 1999, pp. 396–9

111 Berg, Y., and Lande, T. S.: 'Tunable current mirrors for ultra low voltage', Proceedings of the *IEEE International Symposium on Circuits and Systems*, 1999, pp. 17–20

112 Berg, Y., Wisland, D. T., and Lande, T. S.: 'Ultra low-voltage/low-power digital floating-gate circuits', *IEEE Transactions on Circuits System*, 1999 **46**(7), pp. 930–6

113 Berg, Y., and Lande, T. S.: 'Area efficient circuit tuning with floating-gate techniques', Proceedings of the *IEEE International Symposium on Circuits and Systems*, 1999 **2**, pp. 396–9

114 Berg, Y., and Lande, T. S.: 'Ultra low voltage current multiplier/divider', Proceedings of the *IEEE International Conference on Electronics, Circuits and Systems (ICECS)*, 1999, pp. 1369–72

115 Berg, Y., Lande, T. S., and Naess, O.: 'Programming floating-gate circuits with UV-activated conductances', *IEEE Transactions on Circuits and Systems II: Analog and Digital Signal Processing*, 2001 **48**, pp. 12–19

116 Berg, Y., Lande, T. S., Naess, O., and Gundersen, H.: 'Ultra-low-voltage floating-gate transconductance amplifiers', *IEEE Transactions on Circuits and Systems II: Analog and Digital Signal Processing*, 2001 **48**, pp. 37–44

117 Tsividis, Y.: *Operation and modeling of the MOS transistor* (McGraw and Hill, 1987)

118 Foty, D.: *MOSFET modelling with SPICE: principles and practice* (Prentice Hall, 1997)

119 Allen, P. E., and Holberg, D. R.: *CMOS analog circuit design* (Oxford University Press, 2002)

120 *Star-Hspice Manual*, Release 2002.4, Avant! software V2002.4 Copyright

121 Ramírez-Angulo, J., González-Altamirano, G., and Choi, S. C., 'Modeling multiple-input floating-gate transistors for analog signal processing',

Proceedings of the *IEEE International Symposium on Circuits and Systems*, 1997, pp. 2020–3

122 Yin, L., Embabi, S. H. K., and Sánchez-Sinencio, E.: ' A floating-gate MOSFET D/A converter', Proceedings of the *IEEE International Symposium on Circuits and Systems*, 1997, pp. 409–12

123 Tombs, J., Ramírez-Angulo, J., Carvajal, R. G., and Torralba, A.: 'Integration of multiple-input floating-gate transistors into a top-down CAD design flow', *Conference on Design of Circuits and Integrated Systems (DCIS)*, Palma de Mallorca, 1999, pp. 767–71

124 Rodríguez, E., Huertas, G., Avedillo, M. J., Quintana, J. M., and Rueda, A.: 'Practical digital circuit implementations using νMOS threshold gates', *IEEE Transactions on Circuits and Systems II*, 2001 **48**, pp. 102–6

125 Cadence Software

126 Kahng, D., and Sze, S. M.: 'A floating-gate and its application to memory devices', *The Bell System Technical Journal*, 1967 **46**(4), pp. 1288–95

127 Kotani, K., Shibata, T., Imai, M., and Ohmi, T.: 'Clock-controlled neuron-MOS logic gates', *IEEE Transactions on Circuits and Systems-II: Analog and Digital Signal Processing*, 1998 **45**, pp. 518–22

128 Lee, C., Hou, T-H., and Kan, E. C.-C.: 'Nonvolatile memory with a metal nanocrystal/nitride heterogeneous floating-gate', *IEEE Transactions on Electron Devices*, 2005 **52**(12), pp. 2697–702

129 Bandyopadhyay, A., Serrano, G. J., and Hasler, P.: 'Programming analog computational memory elements to 0.2% accuracy over 3.5 decades using a predictive method', Proceedings of the *IEEE International Symposium on Circuits and Systems*, 2005 **3**, pp. 2148–51

130 Rodriguez-Villegas, E., and Barnes, H.: 'Solution to the trapped charge in FGMOS transistors', *IEE Electronic Letters*, 2003 **39**(19), pp. 1416–17

131 Urquidi, C., Ramirez-Angulo, J., Gonzalez-Carvajal, R., and Torralba, A.: 'A new family of low-voltage circuits based on quasi-floating gate transistors', Proceedings of the *IEEE Midwest Symposium on Circuits and Systems. MWSCAS-2002*, 2002 **1**, pp. 93–6

132 Naess, O., Olsen, E. A., Berg, Y., and Lande T.S.: ' A low voltage second order biquad using pseudo floating gate transistors', Proceedings of the *IEEE International Symposium on Circuits and Systems*, 2003, pp. 125–8

133 Ramirez-Angulo, J., Lopez-Martin, A. J., Gonzales Carvajal, R., and Munoz Chavero, F.: 'Very low-voltage analog signal processing based on quasi-floating gate transistors', *IEEE Journal of Solid-State Circuits*, 2004 **39**(3), pp. 434–42

134 Jespers, P. G. A.: *Integrated converters - D to A and A to D Architectures, Analysis and Simulation* (Oxford University Press Inc., New York, 2001)

135 Toumazou, C., Lidgey, F. J., and Haigh, D. G. (eds.): *Analogue IC design: The current mode approach* (IEE Circuits and Systems Series 2, 1989) ISBN: 0 86341 2157

136 Goncalves, R.T., Fiho, S.N., Schneider, M.C., and Galup-Montoro, C.: 'Programmable switched current filters using MOSFET-only current dividers',

Proceedings of the *IEEE Midwest Symposium on Circuits and Systems*, 1995 **2**, pp. 1046–9

137 Cheung, Y.L., and Buchwald, A.: 'A sampled-data switched-current analog 16-tap FIR filter with digitally programmable coefficients in 0.8 μm CMOS', Proceedings of the *IEEE International Solid-State Circuits Conference*, 1997 **429** pp. 54–5

138 Rueda, A., Yufera, A., and Huertas, J.: 'Switched-current wave analogue filters', Chapter 11 in Toumazou, C., Hughes, J. B. C., and Battersby, N. C. (eds.), *Switched-currents: an analogue technique for digital technology*

139 De Lima, J. A., and Cordeiro, A. S.: ' A low-voltage low-power analog memory cell with built-in 4-quadrant multiplication', *IEEE Transactions on Circuits and Systems II: Analog and Digital Signal Processing*, 2003 **50**(4), pp. 191–5

140 de la Vega, A. S., de Queiroz, A. C. M., and Diniz, P. S. R.: 'Adaptive filter implementation using switched-current technique', Proceedings of the *IEEE International Symposium on Circuits and Systems* Vol. I, May 2001, pp. 17–20

141 Mazurek, A., and Wawryn, K.: 'Programable current mode circuits', *The 8th IEEE International Conference on Electronics, Circuits and Systems (ICECS 2001)*, 2001 **2**, pp. 553–6

142 Rodriguez-Vasquez, A., and Espejo, S.: 'Switched-current cellular neural networks for image processing', Chapter 17 in *Switched-currents an analogue technique for digital technology*, Toumazou, C., Hughes, J. B. C., and Battersby, N. C. (eds.)

143 Steyaert, M., Crols, J., and Gogaert, S.: 'Low-voltage analog CMOS filter design' in *Low-voltage/low-power integrated circuits and systems*, Sánchez-Sinencio, E., and Andreou, A. G. (eds.) (IEEE Press Series on Microelectronic Systems, New York, 1999)

144 Sánchez-Sinencio, E.: 'Continuous-time low-voltage current-mode filters' in *Low-voltage/low-power integrated circuits and systems*, Sánchez-Sinencio, E., and Andreou, A. G. (eds.) (IEEE Press Series on Microelectronic Systems, New York, 1999)

145 Schaumann, R., Ghausi, M. S., and Laker, K. R.: *Design of analog filters: passive, active RC and switched capacitor* (Englewood Cliffs, NJ, Prentice-Hall, 1990)

146 Castello, R., Montecchi, F., Rezzi, F., and Baschirotto, A.: 'Low-voltage analog filters', *IEEE Transactions on Circuits and Systems-I*, 1995 **42**(11), pp. 827–40

147 Tsividis, Y. P., and Voorman, J. O. (eds.): *Integrated continuous-time filters: principles, design and applications* (IEEE Press, New York, 1993)

148 Banu, M., and Tsividis, Y.: 'Fully integrated active RC filters in MOS technology', *IEEE Journal Solid-State Circuits*, 1983 **18**, pp. 644–51

149 Tsividis, Y., Banu, M., and Khoury, J.: 'Continuous-time MOSFET-C filters in VLSI', *IEEE Journal of Solid-State Circuits*, 1986 **21**(1), pp. 15–30

150 Park, C. S., and Schaumann, R.: 'Design of a 4-MHz analog integrated CMOS transconductance-C bandpass filter', *IEEE Journal of Solid-State Circuits*, 1988 **23**, pp. 987–96

151 Snelgrove, W. M., and Shoval, A.: 'A balanced 0.9 μm CMOS transconductance-C filter tunable over the VHF range', *IEEE Journal of Solid-State Circuits*, 1992 **27**(3), pp. 314–23

152 Nauta, B.: 'A CMOS transconductance-C filter technique for very high frequencies', *IEEE Journal of Solid-State Circuits*, 1992 **27**(2), pp. 142–53

153 Alini, R., Baschirotto, A., and Castello, R.: 'Tunable BiCMOS continuous-time filter for high-frequency applications', *IEEE Journal of Solid-State Circuits*, 1992 **27**, pp. 1905–15

154 Stefanelli, B., and Kaiser, A.: 'A 2 μm CMOS fifth-order low-pass continuous-time filter for video-frequency applications', *IEEE Journal of Solid-State Circuits*, 1993 **28**(7), pp. 713–18,

155 Rezzi, F., Bietti, I., Cazzaniga, M., and Castello, R.: 'A 70 mW seventh-order filter with 7-50 MHz cutoff frequency and programmable boost and group delay equalization', *IEEE Journal of Solid-State Circuits*, 1997 **32**(12), pp. 1987–99

156 Ganti, R., Carley, L. R., and Myers, B. A.: 'A low distortion high frequency transconductor structure'. Proceedings of the *IEEE International Symposium on Circuits and Systems, II*, 1999 pp. 216–19

157 Pavan, S., Tsividis, Y. P., and Nagaraj, K.: 'Widely programmable high-frequency continuous-time filters in digital CMOS technology', *IEEE Journal of Solid-State Circuits*, 2000 **35**(4), pp. 503–11

158 Giustolisi, G., De, G., and Pennisi, S.: 'High-linear class AB transconductor for high-frequency applications', Proceedings of the *IEEE International Symposium on Circuits and Systems*, 2000, pp. 169–72

159 Voorman, H., and Veenstra, H.: 'Tunable high-frequency Gm-C filters', *IEEE Journal of Solid-State Circuits*, 2000 **35**(8), pp. 1097–108

160 Silva-Martínez, J., Steyaert, M., and Sansen, W.: *High-performance CMOS continuous-time filters* (Kluwer Academic Publishers, 1993)

161 Nedungadi, A., and Viswanathan, T. R.: 'Design of linear CMOS transconductance elements', *IEEE Transactions on Circuits and Systems*, 1984 **31**, pp. 891–4

162 Pennock, J. L.: 'CMOS triode transconductor for continuous-time active integrated filters', *IEE Electronic Letters*, 1985 **21**, pp. 817–18

163 Seevinck, E., and Wasenaar, R. F.: ' A versatile CMOS linear transconductor square-law function circuit', *IEEE Journal of Solid-State Circuits*, 1987 **22**, pp. 366–77

164 Szczepanski, S., Schaumann, R., and Wu, P.: 'Linear transconductors based on cross-coupled CMOS pair', *IEE Electronic Letters*, 1991 **27**, pp. 783–5

165 Tanimoto, H., Koyama, M., and Yoshida, Y.: 'Realization of a 1V active filter using linearization technique employing plurality of emitter-couple pairs', *IEEE Journal of Solid State Circuits*, 1991 **26**, pp. 937–45

166 Silva-Martinez, J., Steyaert, M., and Sansen, W.: ' A large-signal very low-distortion transconductor for high-frequency continuous-time filters', *IEEE Journal of Solid-State Circuits*, 1991 **SC-26**, pp. 946–55

167 Koyama, M., Arai, T., Tanimoto, H., and Yoshida, Y.: 'A 2.5 V active low-pass filter using all n-p-n Gilbert cells with a 1-Vpp linear input range', *IEEE Journal of Solid State Circuits*, 1993 **28**, pp. 1246–53

168 Yang, F., and Enz, C. C.: 'A low distortion BiCMOS seventh-order Bessel filter operating at 2.5 V supply', *IEEE Journal Solid-State Circuits*, 1996 **31**, pp. 321–30

169 Punzenberger, M., and Enz, C.: 'A new 1.2 V BiCMOS log-domain integrator for companding current mode filters'. Proceedings of the *IEEE International Symposium on Circuits Systems*, Vol. 1, Atlanta, GA, 1996, pp. 125–8

170 Punzenberger, M., and Enz, C. C.: 'A compact low-power BiCMOS log-domain filter', *IEEE Journal of Solid-State Circuits*, 1998 **33**(7), pp. 1123–9

171 Szczepanski, S., Jakusz, J., and Schaumann, R.: 'A linear fully balanced CMOS OTA for VHF filtering applications', *IEEE Transactions on Circuits and Systems-II: Analog and Digital Signal Processing*, 1997 **44**(3), pp. 174–87

172 Rodríguez, E. O., Yúfera, A., and Rueda, A.: 'A low-voltage floating-gate MOS transconductor', Proceedings of *XIV Design of Integrated Circuits and Systems Conference*, Palma de Mallorca, Spain, 1999, pp. 539–44

173 Lee, T-S., and Liu, C-C.: 'Design techniques for low-voltage VHF BiCMOS transconductance-C filter with automatic tuning', Proceedings of *International Symposium on Circuits and Systems, II*, 2000, pp. 601–4

174 Rodríguez-Villegas, E. O., Rueda, A., and Yúfera, A.: 'A 1.5 V second-order FGMOS filter', Proceedings of *XV Design of Integrated Circuits and Systems Conf. (DCIS'00)*, 2000, pp. 358–62

175 Rodríguez, E. O., Yúfera, A., and Rueda, A.: 'A G_m-C floating-gate MOS integrator', Proceedings of the *IEEE International Symposium on Circuits and Systems*, 2000, pp. 153–6

176 Rodríguez, E. O., Yúfera, A., and Rueda, A.: 'A low-voltage sqr-root floating-gate Mos integrator', Proceedings of the *IEEE International Symposium on Circuits and Systems*, 2000, pp. 184–7

177 Rodríguez-Villegas, E. O., Rueda, A., and Yúfera A.: 'A 1.5 V second-order FGMOS filter', Proceedings of the *IEEE European Solid-State Circuits Conference*, 2000, pp. 68–71

178 Rodríguez-Villegas, E., Yúfera, A., and Rueda, A.: 'A low-voltage floating-gate MOS biquad', *VLSI Design*, 2001 **12**(3), pp. 407–14

179 Adams, R. W.: 'Filtering in the log domain', Proceedings of *63rd Convention A. E. S.*, May 1979, pp. 1470–6

180 Seevinck, E.: 'Companding current-mode integrator: a new circuit principle for continuous-time monolithic filters', *IEE Electronic Letters*, 1990 **26**(24), pp. 2046–7

181 Tsividis, Y., Gopinathan, V., and Tóth, L.: 'Companding in signal processing', *IEE Electronic Letters*, 1990 **26**, pp. 1331–2,

182 Frey, D. R.: 'Log-domain filtering: an approach to current-mode filtering', *IEE Proceedings-G*, 1993 **140**(6), pp. 406–16

183 Frey, D. R.: 'A general class of current mode filters', Proceedings of the *IEEE International Symposium on Circuits and Systems*, 1993 **2**, pp. 1435–8

184 Frey, D. R.: 'Exponential state filters: a generic current mode design strategy', *IEEE Transactions on Circuits Systems I*, 1996 **43**, pp. 34–42

185 Mulder, J., van der Woerd, A. C., Serdijn, W. A., and Roermund, A. M.: 'Current-mode companding $\sqrt{x}$ domain integrator', *IEE Electronic Letters*, 1996 **32**(3), pp. 198–9

186 Frey, D.: 'Future implications of the log domain paradigm', *IEE Proceedings on Circuits Devices System*, 2000, **147**(1) pp. 65–72

187 Eskiyerli, M., and Payne, A. J.: 'Square root domain filter design and performance', *Analog Integrated Circuits and Signal Processing* 2000 **22**, pp. 231–43

188 López-Martín, A. J., and Carlosena, A.: 'A 1.5 V CMOS square-root domain filter', Proceedings of the *IEEE International Symposium on Circuits and Systems*, 2001, pp. 1465–8

189 Frey, D. R.: 'Log domain filtering for RF applications', *IEEE Journal of Solid State Circuits*, 1996 **31**(10), pp. 1468–75

190 El-Gamal, M., Leung, V., and Roberts, G. W.: 'Balanced log-domain filters for VHF applications', Proceedings of the *IEEE International Symposium on Circuits and Systems*, 1997, pp. 493–6

191 El-Gamal M. N., and Roberts, G. W.: 'Very high-frequency log-domain bandpass filters', *IEEE Transactions on Circuits and System II*, 1998 **45**, pp. 1188–98

192 Payne, A., Khumsat, P., and Thanachayanont, A.: 'Design and implementation of high frequency log-domain filters', Proceedings of the *IEEE International Symposium on Circuits and Systems*, 1999, pp. 972–5

193 El-Gamal, M. N., Baki, R. A., and Bar-Dor, A.: '30-100 MHz NPN-only variable-gain class AB instantaneous companding filters for 1.2 V applications', *IEEE Journal of Solid-State Circuits*, 2000 **35**, pp. 1853–64

194 Serra-Graells, F.: 'VLSI CMOS low-voltage log companding filters', Proceedings of the *IEEE International Symposium on Circuits and Systems*, Vol. I, 2000, pp. 172–5

195 Masmoudi, D., Serdijn, W. A., Mulder, J., van der Woerd, A. C., Tomas, J., and Dom, J. P.: 'A new current-mode synthesis method for dynamic translinear filters and its applications in hearing instruments', *Journal of Analog IC and Signal Processing*, 2000 **22**, pp. 221–9

196 Fried, R., Python, D., and Enz, C. C.: 'Compact log-domain current mode integrator with high transconductance to bias current ratio', *IEE Electronic Letters*, 1996 **32**(11), pp. 952–3

197 Python, D., Punzenberger, M., and Enz, C.: 'A 1 V CMOS log-domain integrator', Proceedings of the *IEEE International Symposium on Circuits and Systems*, Vol. II, 1999, pp. 685–8

198 Python, D., and Enz, C.: 'A micropower class AB CMOS log-domain filter for DECT applications', Proceedings of the *IEEE European Solid-State Circuits Conference*, 2000, pp. 64–7

199 El-Masry, E. I., and Wu, J.: 'Low voltage micropower log-domain filters', *Journal of Analog IC and Signal Processing*, 2000 **22**, pp. 209–20

200 Serra-Graells, F.: 'All-MOS subthreshold log filters', Proceedings of the *International Symposium on Circuits and Systems*, Vol. II, 2001, pp. 137–40

201 Perry, D., and Roberts, G. W.: 'The design of log-domain filters based on the operational simulation of LC ladders', *IEEE Transaction on Circuit and System II: Analog and Digital Signal Processing*, 1996 **43**, pp. 763–74

202 Drakakis, E. M., Payne, A. J., and Toumazou, C.: 'Bernoulli operator: a low level approach to log-domain signal processing', *IEE Electronic Letters*, 1997 **33**(12), pp. 1008–9

203 Fox, R., Nagarajan, M., and Harris, J.: 'Practical design of single-ended log-domain filter circuits', Proceedings of the *IEEE International Symposium on Circuits and Systems*, 1997 **1**, pp. 341–4

204 Himmelbauer, W., and Andreou, A. G.: 'Log-domain circuits in subthreshold MOS', Proceedings of the *IEEE Midwestern Symposium on Circuits and Systems*, 1997 **1**, pp. 26–30

205 López, A., and Carlosena, A.: 'The exponential CCII-, a building block for log-domain circuits', Proceedings of the *IEEE International Conference on Electronics, Circuits, and Systems*, 1998 **2**, pp. 389–92

206 Baswa, S., Bikumandla, M., Ramirez-Angulo, J., Lopez-Martin, A.J., Carvajal, R. G., and Ducoudray-Acevedo, G.: 'Low-voltage low-power super class-AB CMOS op-amp with rail-to-rail input/output swing', Proceedings of the *Fifth IEEE International Caracas Conference on Devices, Circuits and Systems*, 2004 **1**, pp. 83–6

207 Rodriguez-Villegas, E., Yufera, A., and Rueda, A.: 'A 1-V micropower log-domain integrator based on FGMOS transistors operating in weak inversion', *IEEE Journal of Solid-State Circuits*, 2004 **39**(1), pp. 256–9

208 Rodriguez-Villegas, E., Yufera, A., and Rueda, A.: 'A 1.25-V micropower Gm-C filter based on FGMOS transistors operating in weak inversion', *IEEE Journal of Solid-State Circuits*, 2004 **39**(1), pp. 100–11

209 Chawla, R., Haw-Jing Lo, Basu, A., Hasler, P., and Minch, B. A.: 'A fully programmable log-domain bandpass filter using multiple-input translinear elements', Proceedings of the *IEEE International Symposium on Circuits and Systems*, 2004 **1**, pp. I-33 - I-36

210 Rodriguez-Villegas, E.: 'A 0.8 V, 360 nW Gm-C biquad based on FGMOS transistors', Proceedings of the *IEEE International Symposium on Circuits and Systems ISCAS 2005*, May 2005, pp. 2156–9

211 Steyaert, M., and Sansen, W.: 'Power-supply rejection ratio in operational transconductance amplifiers', *IEEE Transaction on Circuits and Systems*, 1990 **37**, pp. 1077–84

212 Johns, D. A., and Martin, K.: *Analog integrated circuit design* (Wiley, New York, 1997)

213 Vittoz, E.: 'Micropower techniques' in *Design of VLSI circuits for telecommunication and signal processing*, Franca, J., and Tsividis, Y. (eds.) (Prentice-Hall, Englewood Cliffs, NJ, 1993)

214 Austria Mikro Systeme International, '0.8 μm CMOS process parameters', 9933006, rev_a, Nov. 1996

215 Rodríguez-Villegas, E. O., Rueda, A., and Yúfera, A., 'A 10.7 MHz FGMOS low power sixth order bandpass filter', Proceedings of *Design Circuits and Integrated Systems Conference* (DCIS), 2001, pp. 68–71

216 Silva-Martinez, J., Steyaert, M. S. J., and Sansen, W.: 'Design techniques for high-performance full-CMOs OTA-RC continuous-time filter', *IEEE Journal of Solid-State Circuits*, 1992 **26**, pp. 993–1001

217 Austria Mikro Systeme International, '0.35 μm CMOS process parameters', 9933016, rev_2.0, Nov. 1998

218 Gilbert, B.: 'Translinear circuits: a proposed classification', *Electronic Letters*, **11**(1), pp. 14–16. See also errata, Vol.11, no.6, pp. 136, 1975

219 Serrano-Gotarredona, T., Linares-Barranco, B., and Andreou, A. G.: 'General translinear principle for subthreshold MOS transistors', *IEEE Transactions on Circuits and Systems I*, 1999 **46**(5), pp. 607–16

220 Shepherd, L., and Townazou, C.: 'Towards direct biochemical analysis with weak inversion ISFET', *IEEE International Workshop on Biomedical Circuits and Systems*, 2004, pp. S1/5–S5-8

221 Yin, Q., Eisenstadt, W. R., and Fox, R. M.: 'A translinear-based RF RMS detector for embedded test', Proceedings of the *2004 International Symposium on Circuits and Systems, 2004. ISCAS '04*, Vol. 1, pp. 245–8

222 Abuelma'atti, M. T.: 'A translinear circuit for sinusoidal frequency division', *IEEE Transactions on Instrumentation and Measurement*, 2005 **54**(1), pp. 10–14

223 Psychalinos, C., Fragoulis, N., and Haritantis, I.: 'Log-domain wave filters', *IEEE Transactions on Circuits and Systems II: Analog and Digital Signal Processing*, 2004 **51**(6), pp. 299–306

224 Baki, R. A., and El-Gamal, M. N.: 'A low-power 5-70-MHz seventh-order log-domain filter with programmable boost, group delay, and gain for hard disk drive applications', *IEEE Journal of Solid-State Circuits*, 2003 **38**(2), pp. 205–15

225 Gerosa, A., Maniero, A., and Neviani, A.: 'A fully integrated dual-channel log-domain programmable preamplifier and filter for an implantable cardiac pacemaker', *IEEE Transactions on Circuits and Systems I: Fundamental Theory and Applications*, 2004 **51**(10), pp. 1916–25

226 Serra-Graells, F.: 'VLSI CMOS subthreshold log companding analog circuit techniques for low-voltage applications', Ph.D. dissertation, Universidad Autónoma Barcelona, 2001

227 Cilingiroglu, U.: 'A purely capacitive synaptic matrix for fixed-weight neural networks', *IEEE Transactions on Circuits and Systems*, 1991 **38**, pp. 210–17

228 Lerch, J. B.: 'Threshold gate circuits employing field-effect transistors', US03715603, Feb. 6, 1973

229 Johnson, M. G.: 'A symmetric CMOS NOR gate for high speed applications', *IEEE Journal of Solid State Circuits*, 1998 **23**, pp. 1233–6

230 Schultz, K. J., Francis, R. J., and Smith, K. C.: 'Ganged CMOS: trading standby power for speed', *IEEE Journal of Solid-State Circuits*, 1990 **25**(3), pp. 870–3

231 Goodwin-Johansson, S.: 'Circuit to perform variable threshold logic', 1990, US048906059

232 Varshavsky, V., 'Beta-driven threshold elements', Proceedings of *Great Lakes Symposium on VLSI*, 1998, pp. 52–8

233 Prange, S., Thewes, R., Wohlrab, E., and Weber, W.: Circuit arrangement for realizing logic elements that can be represented by threshold value equations', 1999, US 05991789

234 Lee, C. L., and Jen, C-W.: 'CMOS threshold gate and networks for order statistic filtering', *IEEE Transactions on Circuits and Systems I: Fundamental Theory and Applications*, 1994 **41**(6), pp. 453–6

235 Quintana, J. M., Avedillo, M. J., Rueda, A., and Baena, C.: 'Practical low-cost CMOS realization of complex logic functions', Proceedings of the *European Conference on Circuit Theory and Design* (ECCTD), 1995, pp. 51–4

236 Goser, K. F., Pacha, C., Kanstein, A., and Rossmann, M. L.: 'Aspects of systems and circuits for nanoelectronics', Proceedings of the *IEEE*, 1997 **85**(4), pp. 558–73

237 Avedillo, M. J., Quintana, J. M., and Rueda, A.: 'Threshold logic', *Wiley encyclopedia of electrical and electronics engineering*, Webster, J. G., (ed.), 1999 **22**, pp. 178–90

238 Rodríguez-Villegas, E., Quintana, J. M., Avedillo, M. J., and Rueda, A.: 'Threshold logic based adders using floating-gate circuits' in *Advances in physics, electronics and signal processing applications* (World Scientific and Engineering Society Press, 2000)

239 Quintana, J. M., Avedillo, M. J., Jiménez, R., Rodríguez Villegas, E., and Rueda, A: 'Improved compressor designs bases on threshold logic', Proceedings of the *Seminario Anual de Automática, Electrónica Industrial e Instrumentación (SAEEI)*, 2001, pp. 1–5 (index)

240 Quintana, J. M., Avedillo, M. J., Rodríguez-Villegas, E., and Rueda, A.: '*v*MOS-based compressor designs', Proceedings of the *12th International Conference on Microelectronics (ICM)*, 2000, pp. 33–6

241 Rodríguez-Villegas, E., Quintana, J. M., Avedillo, M. J., and Rueda, A.: 'Sorting networks implemented as *v*MOS circuits', *IEE Electronic Letters*, 1998 **34**(23), pp. 2237–8

242 Rodríguez-Villegas, E., Avedillo, M. J., Quintana, J. M., Huertas, G., and Rueda, A.: '*v*MOS based sorters for arithmetic applications', *VLSI Design Journal*, 2000 **11**(2), Number j99309

243 Quintana, J. M., Avedillo, M. J., Rodríguez-Villegas, E., Rueda, A., and Huertas, J. L.: 'Efficient *v*MOS realization of threshold voters for self-purging redundancy', Proceedings of the *XIII Symposium on Integrated Circuits and Systems Design*, 2000, pp. 321–6

244 Rodríguez-Villegas, E., Avedillo, M. J., Quintana, J. M., Huertas, G., and Rueda, A.: '*v*MOS-based sorters for multiplier implementations', Proceedings of the *IEEE International Symposium on Circuits and Systems (ISCAS'99)*, Orlando, 1999

245 Zhuang, N., and Wu, H.: 'A new design of the CMOS full adder', *IEEE Journal of Solid-State Circuits*, 1992 **27**(5), pp. 840–4

246 Abu-Khater, I. S., Bellaouar, A., and Elmasry, M. I.: 'Circuit techniques for CMOS low-power high-performance multipliers', *IEEE Journal of Solid-State Circuits*, 1996 **31**(10), pp. 1535–46

247 Gustafsson, O., Dempster, A.G., and Wanhammar, L.: 'Multiplier blocks using carry-save adders', Proceedings of the *IEEE International Symposium on Circuits and Systems*, 2004 **2**, pp. 473–6

248 Quintana, J. M., Avedillo, M. J., Jimenez, R., and Rodriguez-Villegas, E.: 'Low-power logic styles for full-adder circuits'. Proceedings of the *IEEE International Conference on Electronics, Circuits and Systems (ICECS)*, 2001 **3**, pp. 1417–20

249 Hwang, K.: *Computer Arithmetic, Principles, Architecture and Design* (John Wiley & Sons, 1979) ISBN:0471052000

250 Song, P. J., and De Micheli, G.: 'Circuit and architecture trade-offs for high-speed multiplication', *IEEE Journal of Solid-State Circuits*, 1991 **26**(9), pp. 1184–98

251 Zimmermann, R., and Fichtner, W.: Low-power logic styles: CMOS versus pass-transistor logic', *IEEE Journal of Solid-State Circuits*, 1997 **32**(7), pp. 1079–90

252 Lamagna, E. A.: 'The complexity of monotone networks for certain bilinear forms, routing problems, sorting, and merging', *IEEE Transaction on Computing*, 1979 **C-28**, pp. 773–82

253 Batcher, K. E., and Ohmi, T.: 'Sorting networks and their applications'. Proceedings of the *AFIPS Spring Joint Computer Conference*, 1968 **32**, pp. 307–14

254 Leblebici, Y., Özdemir, H., Kepkep, A., and Çilingiroglu, U.: 'A compact (8×8)-bit serial/parallel multiplier based on capacitive threshold logic', Proceedings of the *European Conference on Circuit Theory and Design (ECCTD)*, 1995 pp. 55–8

255 Muroga, S.: *Threshold logic and its applications* (John Wiley & Sons, New York, 1971)

256 Wuu, T.-Y., and Vrudhula, S. B. K.: 'A design of a fast and area efficient multi-input Muller C-element', *IEEE Transactions on VLSI Systems*, 1993 **1**(2), pp. 215–19

Index

Page numbers in *italics* show pages with tables or figures that are separated from text.